Astrophysics and Space Science Proceedings

Volume 60

The Astrophysics and Space Science Proceedings is a series of high-level proceedings volumes covering a broad range of subjects in Astrophysics, Astronomy, Cosmology, and Space Science. The contributors are distinguished specialists with international reputations in their fields of expertise. The subjects are placed in the broader context of related disciplines such as Engineering, Computer Science, Environmental Science, and Nuclear and Particle Physics.

Filomena Bufano • Simone Riggi • Eva Sciacca •
Francesco Schilliro
Editors

Machine Learning for Astrophysics

Proceedings of the ML4Astro International Conference 30 May - 1 Jun 2022

Springer

Editors
Filomena Bufano
Osservatorio Astrofisico di Catania
Istituto Nazionale di Astrofisica
Catania, Italy

Simone Riggi
Osservatorio Astrofisico di Catania
Istituto Nazionale di Astrofisica
Catania, Italy

Eva Sciacca
Osservatorio Astrofisico di Catania
Istituto Nazionale di Astrofisica
Catania, Italy

Francesco Schilliro
Osservatorio Astrofisico di Catania
Istituto Nazionale di Astrofisica
Catania, Italy

ISSN 1570-6591 ISSN 1570-6605 (electronic)
Astrophysics and Space Science Proceedings
ISBN 978-3-031-34166-3 ISBN 978-3-031-34167-0 (eBook)
https://doi.org/10.1007/978-3-031-34167-0

This Springer imprint is published by the registered company Springer Nature Switzerland AG
The registered company address is: Gewerbestrasse 11, 6330 Cham, Switzerland

Paper in this product is recyclable.

Preface

This volume is a collection of original review articles resulting from the invited, oral and poster contributions presented at the 1st International Conference on Machine Learning for Astrophysics (ML4ASTRO), held in Catania from May 30 to June 1, 2022. The conference was organized by the Istituto Nazionale di Astrofisica (INAF), in particular by the Osservatorio Astrofisico di Catania, jointly with the University of Catania, Milano Bicocca and Malta. The principal aim of the conference was to connect researchers actively involved in the field of machine learning and its application to astrophysical science cases.

The astronomical community has indeed entered a new era with the massive growth in data size and complexity generated by current wide-area surveys carried out at different wavelengths, from the radio (e.g. SKA precursors, such as LOFAR, ASKAP and MeerKAT) to optical (e.g. LSST, GAIA). Next-generation surveys are expected to largely exceed the data volume of existing surveys by orders of magnitudes. In this context, machine learning (ML) has now emerged as an invaluable tool to mine the raw data deluge, reducing its complexity and enabling astronomers to produce science-ready data products. The conference has therefore been focused on discussing open issues in different astrophysical fields, how ML techniques were used to tackle them and the relative challenges faced in this process. The conference was also conceived as a place where young researchers and students could have had the opportunity to meet domain experts and, at the same time, do networking with colleagues, thereby promoting an exchange of knowledge of techniques and concepts.

The conference program was split into 12 sessions, including 7 invited talks, 33 contributed talks and 33 poster lightning presentations. Since it was organized during the COVID-19 pandemic period, the organization followed a hybrid mode allowing in-person and remote participations. Invited speakers contributed with a series of review talks on supervised, unsupervised, deep learning and anomaly detection techniques in astronomy, as well as with keynote presentations on time series and transient analysis, innovative devices and software frameworks for ML. A special session was focused on the science program of the Square Kilometre Array and its precursors, and to review the ML activities ongoing within the project.

The conference gained a considerable attention from the community, as it brought together 194 scientists from 34 different countries all over the world. The number of received contributions exceeded by a factor 3.5 the number of allocated talk slots, requiring a hard selection process to be put in place by the Science Organizing Committee (SOC). This was carried out by different referees in blind mode, trying to widen the astrophysical thematic and algorithmic methodologies covered during the three conference days, besides the scientific impact and maturity of the proposed application.

This volume contains a reviewed selection of 41 contributions, capturing the entire spectrum of applications covered during the conference, from more traditional supervised methods (e.g. random forests or decision trees) to deep CNN models for classification, pattern discovery and anomaly detection tasks on multi-channel images or time series data. Each contribution has been peer-reviewed independently by two competed reviewers ensuring the scientific quality level of the manuscript. We are very grateful for the hard work and dedication invested by all participants in the conference, in particular by the students, who have contributed to a very pleasant and friendly atmosphere.

The collection captures a bit of the flavor of the International Conference on Machine Learning for Astrophysics, representing a diverse range of perspectives and approaches and providing insight into the potential of machine learning to revolutionize astrophysical research. We are confident that the ideas and findings presented in these proceedings will inspire future research and foster collaboration among scientists.

Catania, Italy
December 2022

Filomena Bufano
Simone Riggi
Eva Sciacca
Francesco Schilliro

Organization

ML4ASTRO 2022 is organized by the Osservatorio Astrofisico di Catania, Istituto Nazionale di Astrofisica (INAF-OACT), in cooperation with the University of Catania, University of Milano Bicocca and University of Malta.

Scientific Organizing Committee

Filomena Bufano	INAF-OACT, Italy
Simone Riggi	INAF-OACT, Italy
Eva Sciacca	INAF-OACT, Italy
Ugo Becciani	INAF-OACT & University of Catania, Italy
Andrea De Marco	University of Malta, Malta
Giuseppe Vizzari	University of Milano Bicocca, Italy

Local Organizing Committee

Cristobal Bordiu	INAF-OACT, Italy
Fabio Vitello	INAF Istituto di Radioastronomia, Italy
Giulia Miceli	INAF-OACT, Italy
Adriano Ingallinera	INAF-OACT, Italy
Giancarlo Bellassai	INAF-OACT, Italy
Eugenio Martinetti	INAF-OACT, Italy
Valentina Cesare	INAF-OACT, Italy
Federico Incardona	INAF-OACT, Italy
Giuseppe Tudisco	INAF-OACT, Italy
Mario Raciti	INAF-OACT, Italy
Teresa Pulvirenti	INAF-OACT, Italy

Referees

Filomena Bufano	INAF-OACT, Italy
Simone Riggi	INAF-OACT, Italy
Eva Sciacca	INAF-OACT, Italy
Francesco Schilliró	INAF-OACT, Italy
Cristobal Bordiu	INAF-OACT, Italy
Andrea De Marco	University of Malta, Malta
Fabio Vitello	INAF Istituto di Radioastronomia, Italy
Thomas Cecconello	University of Milano Bicocca, Italy
Giuseppe Vizzari	University of Milano Bicocca, Italy

Sponsoring Institutions and Grants

Italian National Institute for Astrophysics (INAF), Italy
Horizon 2020 research and innovation programme of the European Commission under grant agreement No. 863448 (NEANIAS)
INAF PRIN TEC programme 2019 (CIRASA)
MOSAICo projects (Metodologie Open Source per l'Automazione Industriale e delle procedure di CalcOlo in astofisica) funded by Italian MISE, Ministero Sviluppo Economico), "Progetto cofinanziato dall'Unione europea, Fondo europeo di sviluppo regionale—Programma Operativo Nazionale Imprese e Competitività 2014–2020".

Contents

Machine Learning for Hα Emitters Classification

Stavros Akras, Christina Karakotia, and Anna Manou

Abstract Over the last 2 decades, machine learning algorithms have become increasingly popular in Astronomy due to the data provided by several sky-surveys. To explore this large amount of data, it is necessary to apply automatic techniques. Decision tree algorithm is used to develop selection criteria for symbiotic stars, planetary nebulae, Be and Mira candidates. Despite the small samples, our models are characterized by high recall and precision of 0.795/0.949, 0.963/0.908, 0.972/0.934, and 0.850/0.961, after oversampling the minority classes.

1 Introduction

The rapid increase of data in Astronomy in the last two decades, generated by numerous ongoing or completed sky surveys, has led to the significant increase in the application of Machine Learning (ML) algorithms. Some of these algorithms are known as supervised, and they are trained using label data sets to predict some features or classify objects. In astronomy, ML algorithm have been used to explore various problems such as star-galaxy-quasar classifications [1], stellar parameters' prediction [2] among others. The tsunami of new data collected from astronomical sky surveys will never be fully explored by humans following traditional approaches, and thus automatic techniques have become more efficient tools for data mining. One of the most widely used algorithms for classification problems is the decision or classification tree.

S. Akras (✉)
National Observatory of Athens, Athens, Greece

C. Karakotia · A. Manou
Department of Physics, University of Ioannina, Ioannina, Greece

F. Bufano et al. (eds.), *Machine Learning for Astrophysics*, Astrophysics and Space Science Proceedings 60, https://doi.org/10.1007/978-3-031-34167-0_1

2 Decision Tree Approach

Akras and collaborators used the Decision tree algorithm in R aiming to find the colour criteria in the near/mid-infrared that better distinguishes Symbiotic Stars (SySts) [3] and planetary nebulae (PNe) [4] from other types of sources that emit Hα line such as Be stars, Mira stars, young stellar objects, cataclysmic variables, Wolf-Rayet stars, hereafter Hα-emitters.

SySts are long-period interacting binary systems that have a white dwarf and a red giant. So far, up to nearly 300 are known in the Milky Way [5]. PNe represent the final phase of the evolution of low mass stars (1–8 M$\odot$), but only those with diameter<6–8 arcsec are considered for our analysis. Be stars refer to B spectral type stars with effective temperature between 10,000–25,000 K and low emission lines. Mira stars are long-period pulsating stars with rapid mass loss winds.

The training sample was constructed by gathering Hα-emitters from the literature and cross-matching with the 2MASS and AllWISE catalogues [3]. Two lists were created, one that contains only the sources of interest (e.g. SySts) and a second one with all the remaining of Hα-emitters, resulting in a highly imbalanced problem. Here, we present the preliminary results from the application of the decision tree algorithm from the scikit-learn library (v1.1.1) [6] in PYTHON for SySts, PNe, Mira and Be stars. For all the decision tree models, the same training sample was employed from [3].

3 Conclusion

The decision tree algorithm from the scikit-learn library in PYTHON was used to search for the characteristic colour criteria that distinguish different types of Hα-emitters.

Either by duplicating the minority class or employing the SMOTE algorithm, we overcame the imbalance problem of our training dataset, and the preliminary results show an excellent performance for our models with high recall and precision. In particular, we obtained recall and precision of 0.795/0.949, 0.972/0.934, 0.963/0.908 and 0.850/0.961 for SySts, Be, PNe and Mira respectively. These parameters were calculated using the True Positive and True Negative values from the corresponding confusion matrices. Decision tree algorithm will also be applied to young stellar objects, cataclysmic variable and Wolf-Rayet stars.

Figure 1 displays the resulting decision trees for SySts (a), Be stars (b), PNe (c) and Mira stars (d). The most important features for the classification of these specific Hα-emitters are the following colour criteria: J-H, H-K, K-W3, W1-W2, and W2-W3. The new models of SySts and PNe resulted in the same colour criteria like those reported by Akras et al. [3, 4].

Last but not least, random forest algorithm will also be employed in order to further explore the selection criteria of each type of Hα-emitters. Our best selection

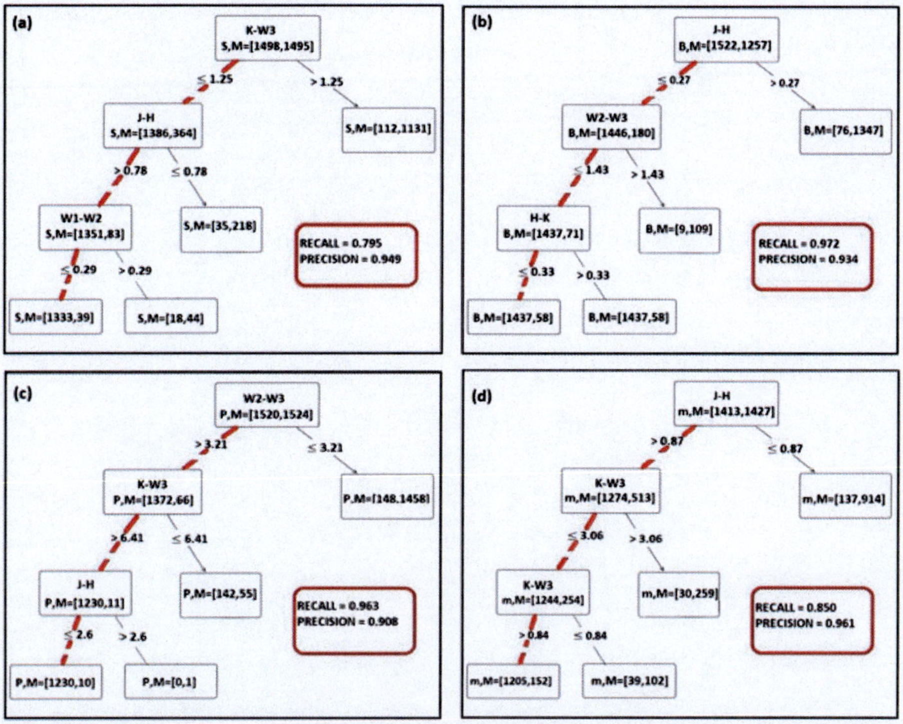

Fig. 1 Decision trees for different Hα emitters. The S, B, P, m and M refer to the numbers of SySts (panel **a**), Be stars (**b**), PNe (**c**), Mira stars (**d**) and mimics, respectively. The red dashed-lines indicate the best colour criteria that distinguish each source from their mimics. The recall and precision of each model are provided in the red boxes

criteria will also be applied to lists of Hα-emitters, aiming to identify new candidates of all the aforementioned types of stars for follow-up spectroscopy.

References

1. Nakazono, L., Mendes de Oliveira, C., Hirata, N.S.T., et al.: Mon. Not. R. Astron. Soc. **507**, 5847 (2021). https://doi.org/10.1093/mnras/stab1835
2. Whitten, D.D., Placco, V.M., Beers, T.C., et al.: Astron. Astrophys. **622**, A182 (2019). https://doi.org/10.1051/0004-6361/201833368
3. Akras, S., Leal-Ferreira, M.L., Guzman-Ramirez, L., et al.: Mon. Not. R. Astron. Soc. **483**, 5077 (2019). https://doi.org/10.1093/mnras/sty3359
4. Akras, S., Guzman-Ramirez, L., & Gonçalves, D.R.: Mon. Not. R. Astron. Soc. **488**, 3238 (2019). https://doi.org/10.1093/mnras/stz1911
5. Akras, S., Guzman-Ramirez, L., Leal-Ferreira, M.L., et al.: Astrophys. J. **240**, 21 (2019). https://doi.org/10.3847/1538-4365/aaf88c
6. Pedregosa, F., Varoquaux, G., Gramfort, A., et al.: J. Mach. Learn. Res. **12**, 2825 (2011)

Stellar Dating Using Chemical Clocks and Bayesian Inference

Andres Moya, Luis Manuel Sarro, Elisa Delgado-Mena, William J. Chaplin, Vardan Adibekyan, and Sergi Blanco-Quaresma

Abstract Dating stars is a major challenge with a deep impact on many astrophysical fields. One of the most promising techniques for this is using chemical abundances. Recent space- and ground-based facilities have improved the number of stars with accurate observations. This has opened the door for using Bayesian inference tools to maximise the information we can extract from them. In this work, we present accurate and reliable stellar age estimates of FGK stars using chemical abundances and stellar parameters, thanks to one of the most flexible Bayesian inference techniques, a hierarchical Bayesian model. The core of the model is a prescription of certain abundance ratios as linear combinations of stellar properties including age. We gathered four different testing sets to assess the accuracy, precision, and limits of our model. With all this, we found that our age estimates and those used as reference agree well. The mean absolute difference of our estimates compared with those used as reference is 0.9 Ga, with a mean difference of 0.01 Ga. When using open clusters, we reached a very good agreement for Hyades, NGC 2632, Ruprecht 147, and IC4651. We also found outliers that are a reflection of chemical peculiarities and/or stars at the limit of the validity ranges of the training set.

A. Moya (✉)
Departament d'Astronomia i Astrofísica, Universitat de València, Burjassot, Spain
e-mail: Andres.moya-bedon@uv.es

L. M. Sarro
Departamento de Inteligencia Artificial, ETSI Informática, UNED, Madrid, Spain

E. Delgado-Mena · V. Adibekyan
Instituto de Astrofísica e Ciências do Espaço, Universidade do Porto, Porto, Portugal

W. J. Chaplin
School of Physics and Astronomy, University of Birmingham, Birmingham, UK

S. Blanco-Quaresma
Harvard-Smithsonian Center for Astrophysics, Cambridge, MA, USA

F. Bufano et al. (eds.), *Machine Learning for Astrophysics*, Astrophysics and Space Science Proceedings 60, https://doi.org/10.1007/978-3-031-34167-0_2

1 Introduction

In this proceeding, we summarize the work of [1] highlighting its most important aspects. Therefore, we refer the reader to that paper for any detail missing in this proceeding. Among all the stellar characteristics, age is one of the most difficult variables to measure because it cannot be directly observed. It must be inferred using diverse methods. One of these techniques is the so-called chemical clocks (CCs) method. It exploits the fact of the chemical evolution of the Galaxy. This chemical evolution is a consequence of the dependence of the thermonuclear reactions operating in stellar interiors (and the atomic elements thereby created) and the stellar evolution depending on the stellar mass.

We can assume that the current stellar surface abundances of certain elements are those of the original cloud from which the star was born, that is, stars act as fossil relics in terms of chemical composition. Therefore, we can use these abundances to estimate the age of the star. This concept has been used in recent years to propose a few chemical abundance ratios for which this evolution is especially clear [2–4, and references therein]. In this paper, we take advantage of the extraordinary data set presented in [3] and the power of machine-learning techniques to go a step further. We train a multi-level or hierarchical Bayesian model combining information from different CCs and stellar effective temperature, metallicity, and gravity to estimate ages. There are two main advantages of using this technique in this context. Firstly, it naturally combines information from different linear regressions. The second advantage is the consistent treatment of uncertainties.

2 Training Data Sample

The data sample we used as the training sample is well described in [1, 3] and references therein. The starting point is a sample of 1059 stars observed within the HARPS-GTO planet search program. All stellar characteristics and needed abundances have been obtained in an homogeneous way. After applying different filters related to the reliability of these determinations, we finally work with a subset of 328 stars.

3 Inference Technique: Hierarchical Bayesian Model

We defined a Bayesian hierarchical model where, in the top layer, the true values of the stellar physical parameters effective temperature $T_{\rm eff}$, iron abundance [Fe/H], (logarithm of the) surface gravity $\log g$, and age t are found. The true values of the abundance ratios used as CCs are (deterministically) modelled as a linear combination of these four stellar parameters,

$$[R_i] = k_{i,0} + k_{i,1} \cdot t + k_{i,2} \cdot T_{\text{eff}} + k_{i,3} \cdot [\text{Fe/H}] + k_{i,4} \cdot \log g\,, \tag{1}$$

where R_i is the i-th abundance ratio used as CC and $k_{i,j}$ is the j-th constant coefficient of the linear combination. The five coefficients $k_{i,j}$ of each linear combination are also model parameters. Finally, observables are set at the lowest level of the model and are defined as random variables normally distributed around the true values and with a standard deviation given by the measurement uncertainties. We used the training set in order to infer the posterior distributions of the $k_{i,j}$ coefficients that were subsequently used to predict ages. Following [3], we used five ratios that involve the Y abundance ([Y/Si], [Y/Mg], [Y/Ti], [Y/Zn], and [Y/Al]) and [Sr/Mg] as CCs because Y is usually easier to obtain than Sr, and its values are also usually more precise. The same model can be used to predict the ages of stars not in the training phase. In this case, we are interested in the posterior probability density of the age given a set of observations of the remaining physical parameters and CCs.

4 Results

To test the HBM constructed with the training sample described in Sect. 2, we gathered four complementary testing samples comprised of stars not belonging to the training set: Twenty-three stars with 'reliable' ages (twenty of them dated using asteroseismology, two belonging to the cluster M67, and the Sun), seventy-nine stars from [5], ten Gaia benchmarks stars, and one hundred and three stars in open clusters. These stars belong to Hyades, NGC 2632, Ruprecht 147, and IC4651.

In Fig. 1 (left panel) we show the comparison between the ages estimated for all the testing field stars with the ages predicted by our HBM. The figure shows that the ages estimated by our model are similar to those provided by the literature. The results are quite stable, despite the fact that the age estimation techniques used for this testing sample are very heterogeneous. Removing the three outliers we obtain an mean absolute difference (MAD) of 0.93 Ga, and an mean difference (MD) of 0.35 Ga in this joint case. For a detailed discussion about the outliers, we refer the reader to [1].

In Fig. 1 (right panel) we provide boxplots of the distribution of estimated ages obtained for four clusters. In addition, we included blue points showing the mean of each distribution, and a brown point for the assumed age of the cluster. The difference between the estimated mean age and the age from the literature is 0.36 Ga, 0.9 Ga, −0.04 Ga, and 0.14 Ga for Hyades, NGC 2632, Ruprecht 147, and IC4651 respectively, with MADs for these clusters of 0.712, 1.35, 1.02, and 0.95 Ga.

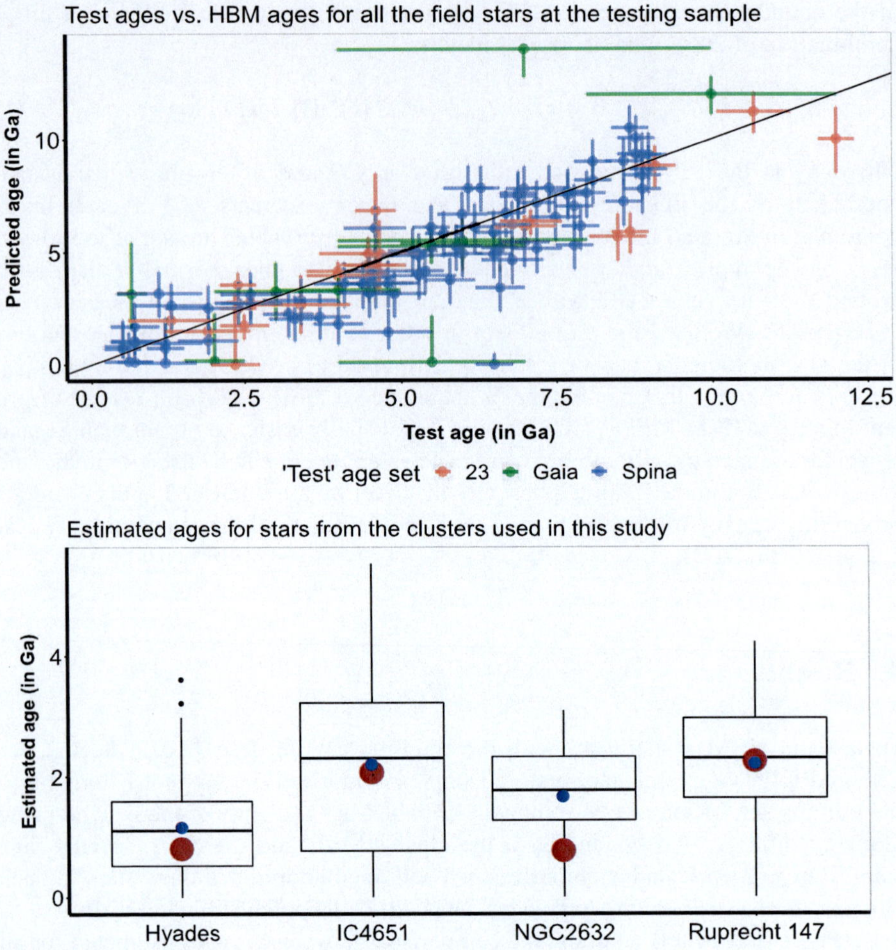

Fig. 1 Age predictions for the testing field stars (left panel) and the testing stars in clusters (right panel) using our HBM. Credit: [1], reproduced with permission, ©ESO

5 Conclusions

We trained a hierarchical Bayesian model using the data set and work of [3] to combine the information coming from different CCs and other physical observables such as T_{eff}, $\log g$, or [Fe/H] to provide robust stellar age estimates with good precision. To test our model, we gathered a number of different testing sets using field stars and stars in clusters. We found that our estimates using CCs and an HBM are similar to the reference ages for almost all the tested stars. Compared with all the testing samples, our estimates present an MAD of about 0.91 Ga, with a really short MD of 0.008 Ga. A most reliable MD was obtained by studying the individual testing subsamples, where MD ranges between −0.4 and 0.4 Ga.

Acknowledgments AM acknowledges funding support from Grant PID2019-107061GB-C65 funded by MCIN/AEI/ 10.13039/501100011033, and from Generalitat Valenciana in the frame of the GenT Project CIDEGENT/2020/036. V.A. and E.D.M. were supported by FCT—Fundação para a Ciência e Tecnologia (FCT) through national funds and by FEDER through COMPETE2020—Programa Operacional Competitividade e Internacionalização by these grants: UID/FIS/04434/2019; UIDB/04434/2020; UIDP/04434/2020; PTDC/FIS-AST/32113/2017 & POCI-01-0145-FEDER-032113; PTDC/FIS-AST/28953/2017 & POCI-01-0145-FEDER-028953, and from FCT through Investigador FCT contracts nr. IF/00650/2015/CP1273/CT0001 and IF/00849/2015/CP1273/CT0003, and POPH/FSE (EC) of the same program.

References

1. Moya, A., Sarro, L.M., Delgado-Mena, E., et al.: Stellar dating using chemical clocks and Bayesian inference. Astron. Astrophys. **660**, A15 (2022). https://doi.org/10.1051/0004-6361/202141125
2. Morel, T., Creevey, O.L., Montalbán, J., et al.: Testing abundance-age relations beyond solar analogues with Kepler LEGACY stars. Astron. Astrophys. **646**, A78 (2021). https://doi.org/10.1051/0004-6361/202039212
3. Delgado Mena, E., Moya, A., et al.: Abundance to age ratios in the HARPS-GTO sample with Gaia DR2. Chemical clocks for a range of [Fe/H]. Astron. Astrophys. **624**, A78 (2019). https://doi.org/10.1051/0004-6361/201834783
4. Viscasillas Vázquez, C., Magrini, L., Casali, G., et al.: The Gaia-ESO survey: Age-chemical-clock relations spatially resolved in the Galactic disc. Astron. Astrophys. **660**, A135 (2022). https://doi.org/10.1051/0004-6361/202142937
5. Spina, L., Meléndez, J., Karakas, A.I., et al.: The temporal evolution of neutron-capture elements in the Galactic discs. Mon. Not. R. Astron. Soc. **474**, 2580 (2018). https://doi.org/10.1093/mnras/stx2938

Detection of Quasi-Periodic Oscillations in Time Series of a Cataclysmic Variable Using Support Vector Machine

Denis Benka, Andrej Dobrotka, and Maximilián Strémy

Abstract Quasiperiodic oscillations (QPO's) in cataclysmic variables (CV) can be very subtle as well as their confidence of detection, indicating less significance than what the reality is. In our observed object, MV Lyrae, we focus on such QPO's. We simulated the QPO according to Timmer and Koenig (Astron. Astrophys. 300:707–710, 1995) and estimated its confidence intervals. Some known (not obvious) QPO's fell under 1-σ and therefore are not significant regarding this method. We propose and evaluate Support Vector Machine (SVM) models trained to identify those QPO's. Our main goal is to testify whether the accuracy of QPO detection using machine learning methods is higher than the significance of confidence levels obtained by the use of Timmer and Koenig's (Astron. Astrophys. 300:707–710, 1995) simulations.

1 Introduction

We analysed the light curve of a nova-like CV, MV Lyrae. These systems show high accretion rate meaning they are in a bright state and good candidates to study accretion processes. We study a phenomenon occurring during the accretion process called flickering which does not have a strict period of occurrence. The QPO's do not produce a single peak but rather a hump i.e., Lorentz profile, meaning the frequencies are not constant and manifest in an interval. We simulated this phenomenon and used the data as a training dataset with two categories. We used a Support Vector Machine (SVM) trained on simulated numerical data with/without the QPO.

D. Benka (✉) · A. Dobrotka · M. Strémy
Advanced Technologies Research Institute, Faculty of Materials Science and Technology in Trnava, Slovak University of Technology in Bratislava, Trnava, Slovakia
e-mail: denis.benka@stuba.sk

F. Bufano et al. (eds.), *Machine Learning for Astrophysics*, Astrophysics and Space Science Proceedings 60, https://doi.org/10.1007/978-3-031-34167-0_3

2 Methodology

We analysed optical Kepler data of MV Lyrae, a nova-like system. We extracted a 272-day long light curve of the object. First, we divided the light curve into 10-day subsamples. Using Lomb-Scargle [1] (L-S) algorithm we subsequently calculated a periodogram for each subsample. The periodograms had a log-log scale and the main reason for this was to identify potential QPO's. We constructed power density spectra (PDS) to see the variability more clearly. Bartlett's [2] method was used to average the periodograms. The PDS had a binning of 0.05. Our target QPO at $\log(f/\mathrm{Hz}) \simeq -3.4$ is defined only with three bins (Fig. 1). As a next step we fitted the PDS with a Lorentz model consisting of four (or three) components. Goodness of the fit using reduced chi-square was $\chi^2 = 0.99$.

The fitting parameters were used as an input to create simulations of the observed QPO using the Timmer and Koenig [3] method. We then estimated confidence intervals based on the simulations. The simulated PDS in numerical form were used as a training dataset for our SVM. Each PDS was represented by a 2D matrix (42×2) of frequency and the power of the signal in log scale. Several dataset sizes were used, i.e. 50, 100, 250, 500, 1000. We also used a validation set (20% of training data) as well as k-fold cross-validation during the training. All of the SVM's used a radial basis function (RBF) kernel, $e^{\gamma \|x-x'\|^2}$, where the hyperparameter γ was optimised using a grid search as well as the margin parameter C of the support vectors of the SVM.

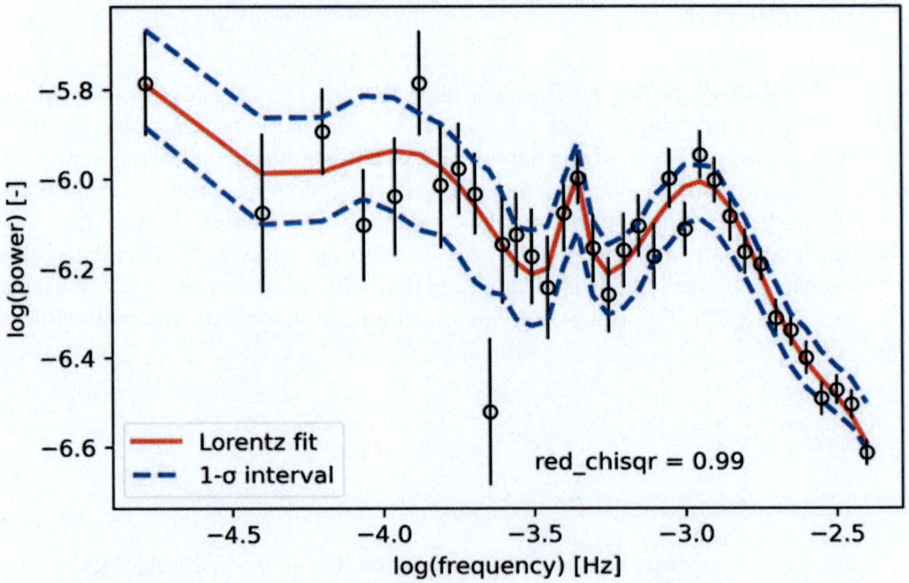

Fig. 1 Fitted simulated PDS and 1-σ intervals. Red is the four component Lorentz fit, blue is the 1-σ interval and PDS bins are black. Source: own research

Table 1 Testing metrics of our SVM's using real data and k-fold cross validation. Source: own research

Dataset size	50	100	250	500	1000
Accuracy	0.54	0.45	0.54	0.54	0.46
Precision	0.55	0.47	0.52	0.53	0.45
Recall	0.53	0.46	0.53	0.54	0.48
F1-score	0.54	0.46	0.52	0.53	0.46

3 Results, Discussion and Conclusion

For testing purposes, 26 real observational data were used out of which 12 show an obvious QPO. Training accuracy was approx. 60% in all cases. The resulting testing accuracies using k-fold cross validation are shown in Table 1. Mean testing accuracy was 51%. In comparison to the confidence levels of Timmer and Koenig [3], our approach did not improve the confidence of QPO detection. The rate of positive detections, or recall of our SVM's was higher while using larger dataset. Increasing the number of training samples also slightly improved the precision, or the true positive rate.

Using 10-fold cross-validation increased the testing accuracy by approx. 1%. We tried using a 5000-sample training dataset. The training process took 2 weeks to complete and the results were even worse than the former (42% testing accuracy using real data). In our case we should have focused more on smaller number of training samples as large datasets proved to have worse metrics. Truth is, the data in each target class overlap in a sense that the features have similar properties. This results in falling in unwanted local optima due to the nature of the optimisation kernel used. This could cause problems for our low-bias RBF kernel to distinguish the ground truth and noise. We also tried to use a fourth degree polynomial kernel. The results were not plausible as the testing accuracy was merely above 30%. In our future work we would like to tailor our SVM using a training dataset with 50 samples while focusing on the preprocession of the data. We did not preprocess our training data as we were afraid it could bring noise to our data. We also think about changing the training data type to image.

Acknowledgments The authors gratefully acknowledge the contribution of the Scientific Grant Agency of the Slovak Republic under the VEGA grant 1/0408/20, and the European Regional Development Fund, project No. ITMS2014+: 313011W085.

References

1. Lomb, N.: Least-squares frequency analysis of unequally spaced data. Astrophys. Space Sci. **39**, 447–462 (1976)
2. Bartlett, M.S.: On the theoretical specification and sampling properties of autocorrelated time-series. Suppl. J. R. Stat. Soc. **8**, 27–41 (1946). https://doi.org/10.2307/2983611
3. Timmer, J., Koenig, M.: On generating power law noise. Astron. Astrophys. **300**, 707–710 (1995)

Dust Extinction from RF Regression of Interstellar Lines

Santiago González-Gaitán and Claudia P. Gutiérrez

Abstract Dust extinction is ubiquitous in the Universe and a challenge in the correction of the brightness and color in astronomical observations. Spectral absorption lines from abundant gas atoms in the interstellar medium (ISM) like sodium, potassium and calcium, or molecules like diffuse interstellar bands, among others, serve as dust indicators and have been used to estimate dust extinction. However, several caveats and limitations exist like line saturation at high optical depths. We explore here new automated avenues to infer dust extinction from spectral lines with a large sample of supernova spectra. We first develop an automated technique to measure equivalent widths with high accuracy when compared to high-resolution spectra and simulated spectra. Secondly, with published Milky Way extinction values (Schlafly and Finkbeiner, Astrophys. J. 737:103, 2011), we train a random forest regression that uses all available lines as input features finding much better predictions than traditional methods based on trends from measurements of single species.

1 Introduction

The interstellar medium (ISM) plays a fundamental role in the galactic enrichment of the galaxies and the universe. It is composed of dust and gas and its presence was inferred from spectral lines at least a century ago [1]. A relation between the strength of these lines and the distance to the Milky Way (MW) stars was recognized in early studies [2], which ultimately was an indication of the relation between the line strength and the amount of dust. Later on, interstellar lines were also seen in extragalactic sources including supernovae (SN) [3]. Nowadays, they are routinely

S. González-Gaitán (✉)
CENTRA, Instituto Superior Técnico, Universidade de Lisboa, Lisboa, Portugal

C. P. Gutiérrez
Finnish Centre for Astronomy with ESO (FINCA), University of Turku, Turku, Finland
e-mail: claudia.gutierrez@utu.fi

F. Bufano et al. (eds.), *Machine Learning for Astrophysics*, Astrophysics and Space Science Proceedings 60, https://doi.org/10.1007/978-3-031-34167-0_4

detected and their strengths are known to correlate with dust extinction [4], such that various empirical relations have been suggested [5–7]. These relations are for the strongest of these interstellar lines, the sodium doublet Na I D. However, the universality of the method is hindered as the line saturates, particularly at low-resolution spectra [8]. Other insterstellar lines like potassium, K I, and the unknown diffuse interstellar bands (DIBs) at 5780Å have shown to be also potential good tracers of dust [9, 10].

In this contribution, instead of using a single species, we aim to simultaneously take advantage of multiple insterstellar lines as a diagnostic for dust extinction in the MW from a large sample of supernova spectra from the literature [e.g. [11, 12]]. For this we employ an automated technique to measure equivalent widths (see Sect. 2.1) and a random forest regression (see Sect. 2.2) to obtain the MW values of [13] (see Sect. 3).

2 Methodology

2.1 Automated Equivalent width Measurement

Based on [14], we develop an automated technique that finds a continuum around the insterstellar line by smoothing the spectrum with a cosine kernel. The nodes for the continuum are evenly spaced in velocity space and the optimal distance between nodes is obtained by maximizing the signal-to-noise of the equivalent width (EW). The EW is obtained by integrating the ratio of the spectrum flux, f, to the continuum, c, over a fixed wavelength window of 2000 km/s: EW=$\sum_i 1 - f_i/c_i$, where i represents each wavelength. When multiple spectra per supernova are available, we first stack each ratio, f/c, before the integration (see Fig. 1). We do not take into account spectra that have too low signal-to-noise or with a continuum slope that is too steep, since this affects the measurement. The uncertainties on the measurement are obtained from a 1σ of the distribution of bootstrap simulations where different individual spectra are randomly used to obtain the EW. The technique is robust for low-resolution spectra as comparisons with high-resolution and simulations show.

2.2 Random Forest Regression

To fully exploit the entire information from the interstellar lines of supernova spectra, we use a Random Forest (RF) Regressor [15, 16] that receives as input the automated EW measurements of following species: Na I D, K I $\lambda\lambda$7665Å, K I $\lambda\lambda$7699Å, Ca II H, Ca II K, DIB $\lambda\lambda$5780Å and DIB $\lambda\lambda$4428Å, with their respective uncertainties, as well as the number of spectra used in the automated measurement

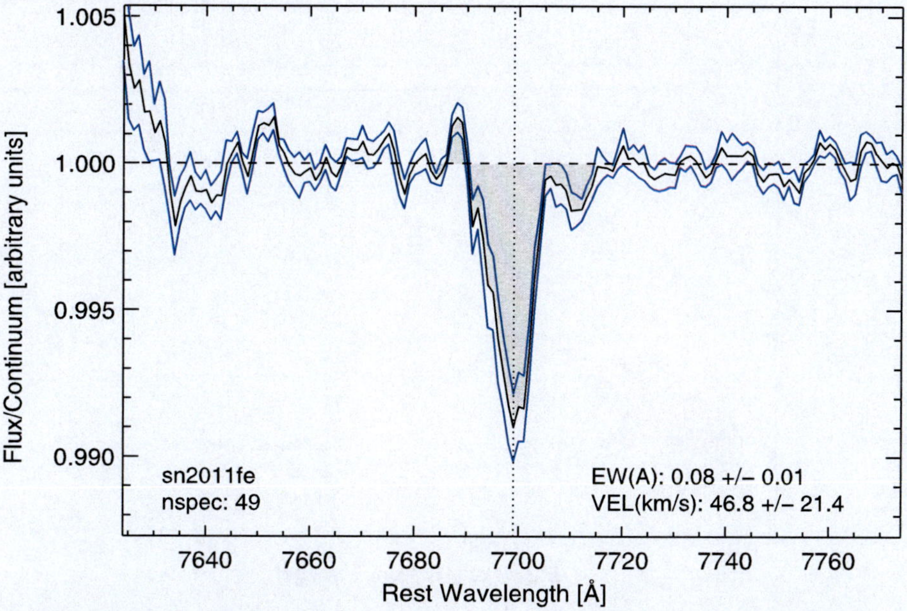

Fig. 1 Stacked flux-to-continuum spectra of SN 2011fe (black) centered around the interstellar line of K I λλ7699Å. The blue lines correspond to the 1σ uncertainties from a bootstrap simulation and the shaded area the integrated EW

and the SN redshift. A total thus of 16 features with a target of continuous extinction values, $E(B-V)$, from the MW [13]. The cost function is the median absolute error, MAE$=\sum_j |EBV_j^{\mathrm{pred}} - EBV_j^{\mathrm{MW}}|/n$, where the sum goes over all n SNe in the sample. 75% of the sample is used as training with a cross-validation changing the training sets ($k=10$) in order to find the optimal hyperparameters such as the number of trees and the maximum depth of the tree. To take uncertainties of the EW measures into account, we do a Monte Carlo of 100 simulations where each EW value is shifted within its uncertainties, and the prediction re-evaluated. The final prediction and uncertainties come from the median and deviation of all realizations.

3 Results

Figure 2 shows the result of the predicted MW values vs the expected ones. Compared to the traditional analytic linear prediciton from a single species (sodium), the median MAE from the RF is already an order of magnitude better. This shows the inmense potential of machine learning techniques and the benefits of using multiple lines. However, the results are far from satisfactory and this is mostly due to the low number of objects used in the training. Our original sample of $\sim$ 1000 SNe gets

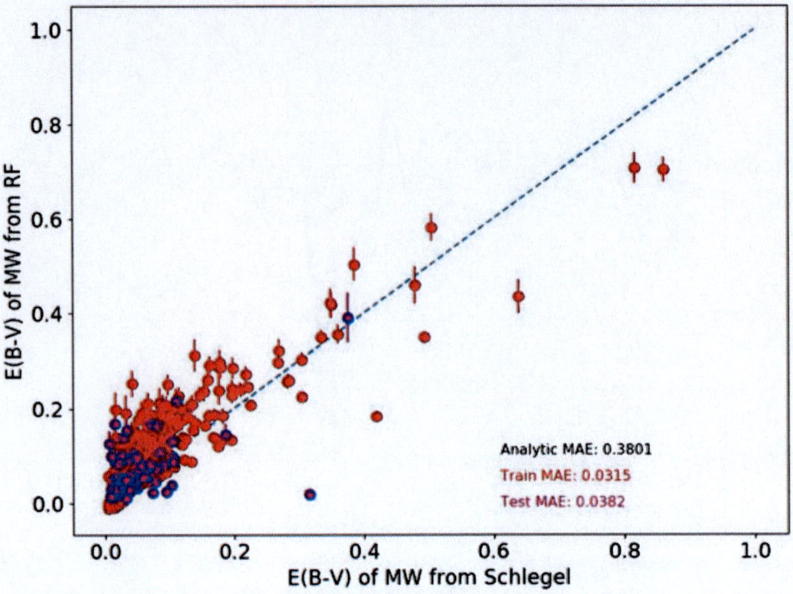

Fig. 2 Reddening values predicted by the RF of multiple interstellar line species vs the values of the MW from [13]. Purple points show the test set and the red points the training set. Diagonal line is a 1:1 relation

diminished to less than 100 when using all interstellar lines. If we use less lines, we retain few hundreds of SNe but the results are anyhow similar. We therefore aim to apply the same methodology to a much larger sample of extragalactic sources from the Sloan Digital Sky Survey that includes almost a million galaxies and quasars, providing much better statistics to explore this better.

References

1. Heger, M.L.: Stationary sodium lines in spectroscopic binaries. Publ. Astron. Soc. Pac. **31**, 304 (1919)
2. Struve, O.: Further work on interstellar calcium. Astrophys. J. **67**, 353 (1928)
3. Penston, M.V., Blades, J.C.: Interstellar NA I and CA II absorption in the spectrum of the recent supernova in NGC 4321. Mon. Not. R. Astron. Soc. **190**, 51 (1980)
4. Buscombe, W., Kennedy, P.M.: Optical interstellar absorption features. Mon. Not. R. Astron. Soc. **139**, 417 (1968)
5. Munari, U., Zwitter, T.: Equivalent width of NA I and K I lines and reddening. Astron. Astrophys. **318**, 269 (1997)
6. Turatto, M., et al.: Variety in supernovae. In: From Twilight to Highlight: The Physics of Supernovae: Proceedings of the ESO/MPA/MPE Workshop Held at Garching, Germany, p. 200. Springer-Verlag (2003)
7. Poznanski, D., et al.: An empirical relation between sodium absorption and dust extinction. Mon. Not. R. Astron. Soc. **426**, 1465 (2012)

8. Poznanski, D., et al.: Low-resolution sodium D absorption is a bad proxy for extinction. Mon. Not. R. Astron. Soc. **415**, L81 (2011)
9. Phillips, M.M., et al.: On the source of the dust extinction in type Ia supernovae and the discovery of anomalously strong Na I absorption. Astrophys. J. **779**, 38 (2013)
10. Kos, J., Zwitter, T.: Properties of diffuse interstellar bands at different physical conditions of the interstellar medium. Astrophys. J. **774**, 72 (2013)
11. Silverman, J.M., et al.: Berkeley supernova Ia program - I. observations, data reduction and spectroscopic sample of 582 low-redshift Type Ia supernovae. Mon. Not. R. Astron. Soc. **425**, 1789 (2012)
12. Blondin, S., et al.: The spectroscopic diversity of type Ia supernovae. Astrophys. J. **143**, 126 (2012)
13. Schlafly, E.F., Finkbeiner, D.P.: Measuring reddening with Sloan digital sky survey stellar spectra and recalibrating SFD. Astrophys. J. **737**, 103 (2011)
14. Förster, F., et al.: Evidence for asymmetric distribution of circumstellar material around type Ia supernovae. Astrophys. J. Lett. **754**, L21 (2012)
15. Ho, T.K.: Random decision forests. In: Proceedings of 3rd International Conference on Document Analysis and Recognition, vol. 1, p. 278 (1995)
16. Pedregosa, F., et al.: Scikit-learn: machine learning in python. J. Mach. Learn. Res. **12**, 2825 (2011)

QSOs Selection in Highly Unbalanced Photometric Datasets: The "Michelangelo" Reverse-Selection Method

Giorgio Calderone, Francesco Guarneri, Matteo Porru, and Stefano Cristiani

Abstract We present a novel selection method aimed at efficiently identifying high-redshift QSOs in highly unbalanced photometric datasets. The method relies on a gradient boosting algorithm, although it may be be used with any other machine learning method providing classification probabilities. We applied the selection method on a photometric dataset and compared its performances to its direct-selection method counterpart, showing that the former privileges the completeness, while the latter privileges the success rate.

1 Introduction

The absorption lines superimposed to the spectra of high-redshift quasi-stellar objects (QSOs) provides valuable information about the distribution, density, temperature and ionization status of the Inter-Galactic Medium (IGM), which in turn is a fundamental tool for cosmology studies concerning the history of the Universe, its dynamical evolution, fundamental physics, etc. Hence, the availability of bright and distant QSOs acting as cosmological beacons to probe the IGM is a first necessary condition to carry out such investigations.

Historically, the most complete spectroscopical surveys to identify such high redshift QSOs had been carried out mainly in the Northern Hemisphere (e.g. SDSS). However, today's most advanced spectrographs facilities, often attached to telescopes with very large collecting area, are located in the South (e.g. ESPRESSO, UVES and, in the near future, ELT/ANDES). Therefore we need to fill the gap, and identify the brightest and highest redshift QSOs in the South, to exploit the capabilities of such observing facilities. On the other hand, photometric surveys are

G. Calderone (✉) · S. Cristiani
INAF-Osservatorio Astronomico di Trieste, Trieste, Italy
e-mail: giorgio.calderone@inaf.it

F. Guarneri · M. Porru
Dipartimento di Fisica, Sezione di Astronomia, Università di Trieste, Trieste, Italy

F. Bufano et al. (eds.), *Machine Learning for Astrophysics*, Astrophysics and Space Science Proceedings 60, https://doi.org/10.1007/978-3-031-34167-0_5

available for the entire sky hence the QSOs in the South had likely been already observed, they simply miss a spectroscopic classification. The QUBRICS[1] survey aims to identify such bright and high-z QSOs in the Southern Hemisphere, using data from available photometric surveys and selection methods based on machine learning. The surveys started in 2018 and collected so far ~ 900 spectra, identifying 642 new QSOs with $z > 2.5$ and $i < 18.5$, with a success rate of $\sim 70\%$ [1, 2]. The current purpose of the survey is to identify most of the bright high-z QSOs in the photometric datasets (Sect. 2), hence we wish to maximize as much as the possible the selection *recall*,[2] while keeping the *precision*[3] at acceptable levels.

In this paper we present a method (Sect. 3) to boost the recall in selection methods, even in presence of a highly imbalanced training dataset (i.e. a dataset where the number of interesting sources, namely high-z QSOs is significantly lower than other sources, such as stars).

2 Data Sets

To test the performance of our selection method we prepared a photometric dataset starting from the PanSTARRS1 (DR2) survey with declination $< 15°$, galactic latitude $> 25°$ (in absolute value) and Y band PSF magnitude $14 < Y < 18.5$. We cross-matched the resulting sources with both the Gaia (EDR3) and the AllWise catalogs, and collected all available magnitudes. Finally, we cross-matched our dataset with our internal QUBRICS database as well as other catalogs from literature, in order to assign a spectroscopic classification to the sources (when available). The composition of our *main sample* is shown in Table 1. As anticipated, the interesting sources (high-z QSOs) account for only 0.006% of the total sample. We split the main sample in a training (80%) and test (20%) sample, following a stratified approach (i.e. class ratios will be kept approximately constant in both the training and test datasets).

Table 1 Composition of the main sample used in this work

Class	No. of sources	Fraction (%)
Stars	27,985,912	90.875
Inactive galaxies	150,581	0.489
Low-z QSOs and AGNs ($z < 2.5$)	16,262	0.053
High-z QSOs ($z > 2.5$)	1847	0.006

[1] QUasars as BRIght beacons for Cosmology in the Southern hemisphere, https://qubrics.oas.inaf.it/.

[2] Ratio of true positives and total number of high-z QSOs in the photometric datasets.

[3] Ratio of true positives to the total number of high-z QSO candidates.

3 QSO Selection

Several machine learning frameworks provide facilities to address multi-label classification problems. However, the class imbalance of the dataset (Table 1) may be detrimental for performances [3], as well as being a possible source of bias for the performance estimators themselves [4]. Moreover, it is not easy to target the recall as the metric to maximize, rather than precision , by simply varying the algorithm hyper-parameters of such *direct selection* methods.

Here we propose our *reverse selection* method to boost recall in presence of highly imbalanced datasets, by exploiting the capabilities of any classifier able to estimate the probability for a source to belong to belong to a class. The steps are as follows:

- Start with the class with largest number of sources (stars in our case), and associate a *non-star* label to all the remaining sources;
- Train a binary classifier to distinguish a *star* from a *non-star*, and predict classifications on both the training and test datasets. If the predicted label for a source is *star* with a probability greater than a threshold τ the source is discarded. In all other cases we discard the prediction but keep the source. The datasets are thus automatically rebalanced towards the less common and most interesting sources (high-z QSOs);
- Repeat the two steps above for the next most abundant classes, i.e inactive galaxies and low-z QSOs (high-z QSOs will be handled at the last step), in place of stars;
- As a last step apply a multi-label classifier (with no probability threshold) to identify the classification on all sources still present in the dataset.

To calculate the precision and recall in the test dataset we considered the predicted label identified in the last step or, if the source has been discarded, the label which has been associated with probability higher than τ. To compare the performances of the direct and reverse selection methods we ran 100 analysis by randomly selecting the training vs. test. split, and used the XGBoost [5] framework for classification (with $\tau = 0.8$ and the same hyper-parameters for both methods). The resulting histograms for precision and recall are shown in Fig. 1. Our results shown that the reverse selection method significantly boosts the recall for the minority class (namely, high-z QSO) at the expense of a small decrease in precision. Also, if we consider the recall in the subsample of QSOs with $z > 3$, i.e. for a redshift slightly larger than the threshold used to distinguish low- from high-z labels, the boost in recall is even more significant.

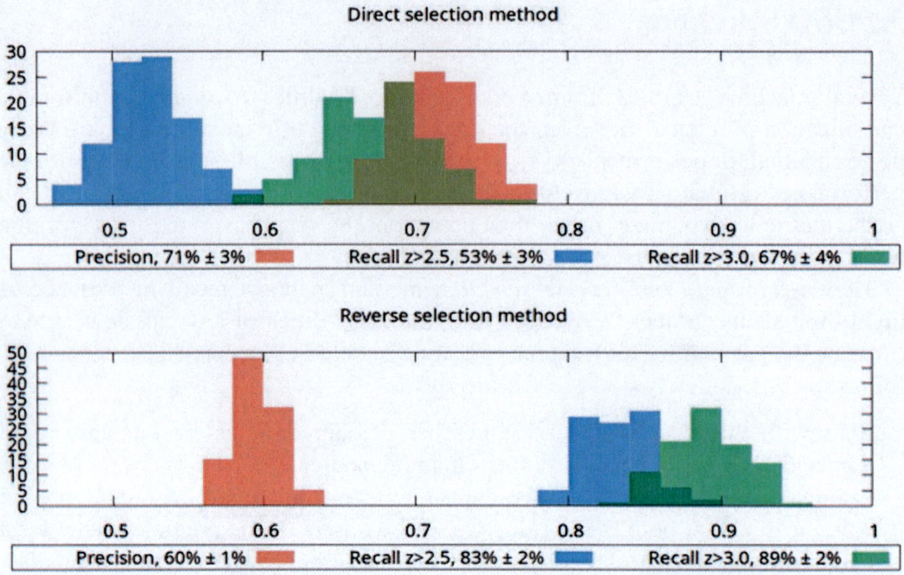

Fig. 1 Distributions of precision (red) and recall (blue for QSOs with $z > 2.5$, green for $z > 3$) histogram for the direct (upper panel) and reverse (lower panel) selection methods

References

1. Calderone, G., et al.: Finding the brightest cosmic beacons in the southern hemisphere. Astrophys. J. **887**, 268 (2019)
2. Guarneri, F., et al.: The probabilistic random forest applied to the selection of quasar candidates in the QUBRICS survey. Mon. Not. R. Astron. Soc. **506**, 247 (2021)
3. Prati, R., et al.: Data mining with imbalanced class distributions: Concepts and methods. In: IICAI (2009)
4. Batista, G., et al.: A study of the behavior of several methods for balancing machine learning training data. In: SIGKDD Explorations, vol. 6, p. 20 (2004)
5. Chen, T., Guestrin, C.: XGBoost: a scalable tree boosting system. In: Proceedings of the 22nd ACM SIGKDD Conference, New York (2016)

Radio Galaxy Detection Prediction with Ensemble Machine Learning

Rodrigo Carvajal, Israel Matute, José Afonso, Stergios Amarantidis, and Davi Barbosa

Abstract The study of Active Galactic Nuclei (AGN) is fundamental to comprehend their evolution and connection with star-formation history and galaxy evolution. Powerful radio emission from AGN traces the largest structures in the Universe and, given the radiative and kinetic energy associated with the phenomena, is a prime feedback candidate to understand the correlations between the properties of the Super-massive Black Hole (SMBH) and the host galaxy. The Epoch of Reionisation (EoR) witnessed the birth of the first luminous sources and the initial steps of the SMBH-host connection. But few AGN have been identified in the EoR of which only a small fraction have radio detections. Recent and future large-scale surveys render the use of regular AGN detection and redshift estimation techniques inefficient. On the other hand, Machine Learning (ML) methodologies can help overcome the computational bottleneck to predict the presence of an ever-increasing number of AGN up to the highest redshifts.

We have developed a series of ML models that, using multi-band photometry, select a list of candidates with their predicted redshift. Models were trained and tested on the The Hobby-Eberly Telescope Dark Energy Experiment (HETDEX) and Stripe 82 fields. We find that AGN selection and redshift estimation metrics are similar to traditional techniques but with a fraction of the computational cost. The pipeline recovers 50%–60% of the radio population, $2\times - 5\times$ better than chance selection, allowing to shed some light on the origin and duty cycle of radio emission.

R. Carvajal (✉) · I. Matute · J. Afonso · D. Barbosa
Instituto de Astrofísica e Ciências do Espaço, Universidade de Lisboa, OAL, Tapada da Ajuda, Lisbon, Portugal

Departamento de Física, Faculdade de Ciências, Universidade de Lisboa, Lisbon, Portugal
e-mail: rcarvajal@alunos.fc.ul.pt

S. Amarantidis
Instituto de Astrofísica e Ciências do Espaço, Universidade de Lisboa, OAL, Tapada da Ajuda, Lisbon, Portugal

Institut de Radioastronomie Millimétrique (IRAM), Granada, Spain

F. Bufano et al. (eds.), *Machine Learning for Astrophysics*, Astrophysics and Space Science Proceedings 60, https://doi.org/10.1007/978-3-031-34167-0_6

1 Overview

Powerful radio emission from Active Galactic Nuclei (AGN) is a prime tracer of active accretion into a Super-massive Black Hole (SMBH). The reasons for its triggering (occurring for ∼5%–30% of AGNs), are still highly uncertain. This radio emission is thought to be one of the mechanisms responsible for the observed correlation between the properties of the host galaxy and the SMBH. Confirming their nature requires detailed knowledge of their demographics up to the highest redshifts which we are still lacking.

In addition, the advent of recent and future surveys with next-generation observatories will render traditional analysis techniques difficult to apply at such large scale (e.g. Evolutionary Map of the Universe, EMU [15]). Advancing our knowledge of the field requires therefore alternative and efficient approaches to identify these sources at all redshift. Machine Learning (ML) provides the tools to understand very large volumes of data without extremely long computing time.

With this goal in mind, we have created a pipeline based on ML techniques to predict AGN at radio wavelengths and their redshift value using only photometric information from non-radio bands.

2 Methodology

We have gathered multi-wavelength photometric data for ∼6.7 million Wide-field Infrared Survey Explorer (WISE) [19] detected sources from the CatWISE2020 catalogue [14] over 424 deg^2 in the Hobby-Eberly Telescope Dark Energy Experiment (HETDEX) Spring field [10]. These photometric data (cross-matched with Panoramic Survey Telescope and Rapid Response System [Pan-STARRS, 8] DR1, Two Micron All-Sky Survey [2MASS, 3, 18], and AllWISE [4] surveys) were used for training, testing and validation of the models and the prediction pipeline, On the other side, we have collected, in the same fashion as with HETDEX, multi-wavelength data from ∼370,000 CatWISE2020-detected sources from ∼92 deg^2 in the Stripe 82 field (S82), which serves as an alternative testing field. Known spectroscopic labels (AGN-galaxy classification) and redshifts for the main catalogue are provided by the Million Quasar Catalog (MQC, [7]) and The Sloan Digital Sky Survey (SDSS)-DR16 [1] for 83,409 sources in the HETDEX field and for 3304 sources in S82. The radio photometric information was obtained by cross-matching the main catalogue with several radio surveys [5, 9, 11, 12, 17] and a flag was created displaying whether or not a source shows a detection in, at least, one of the selected radio surveys. Only sources with spectroscopic classification and redshift were used for training, testing, and validation steps. A 20% of the spectroscopically-confirmed data was used as validation sub-set. From the remaining sources, 80% of them were used for training and 20% for testing.

The final pipeline is composed of three different instances executed in sequence: (1) AGN-galaxy classification; (2) Radio detection prediction and (3) redshift estimate. Following the results presented by Carvajal et al. [2], we have used an ensemble of models for the training stages. Each instance finds the best models (meta learner) to stack the results from the remaining algorithms (base learners) as well as the best subset of features to use (from a total of 66 possibilities, including magnitudes as colours). For each prediction step, five tree-based models were tested: `CatBoost`, `XGBoost`, `ExtraTrees`, `Gradient-Boosting`, and `RandomForest`. For the AGN-galaxy classification and radio-detection models, the selected meta learner is `CatBoost`. In the case of the redshift value prediction, the meta model is `RandomForest`. The remaining models are used as base learners.

3 Results

Metrics based on the pipeline results (F1 score, Matthews Correlation Coefficient, and Recall), as well as the confusion matrix and true-predicted redshift comparison, are presented in Fig. 1. Despite the heterogeneity of the data, the AGN selection and redshift determination metrics are in line with recent results in the literature based in more homogeneous datasets. Our main finding is the significant improvement in the selection of radio AGN sources (more than 52% of recall) which is 2–5× better than a random selection of radio sources from an AGN sample (the fraction of radio-loud AGN is in the range of 5%–30%) [6, 13, 16]. In addition, we find that the metrics of the pipeline are consistent when applied to a non-trained field (such as S82).

Field	F1	MCC	Recall
HETDEX	0.5255	0.4733	0.5235
Stripe 82	0.5395	0.3554	0.5386

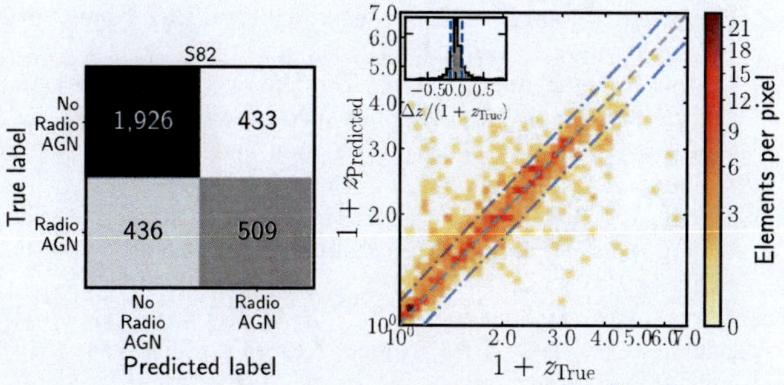

Fig. 1 Summary of results from predictions. (*Left*) Overall metrics of the pipeline applied to the HETDEX and S82 fields: F1-score, Matthews Correlation Coefficient (MCC), and Recall. (*Centre*) Confusion matrix for radio detection prediction on S82. (*Right*) Comparison between true redshifts and those predicted by our pipeline over the sources predicted to be radio AGN in S82. Outliers lie outside the dashed-dotted line. Inset shows the dispersion between true and predicted values with a mean value of $\langle \frac{\Delta z}{1+z_{\mathrm{True}}} \rangle = 0.0768$

In addition, and when applied to non-labelled CatWISE2020 detections, the pipeline predicts an additional 99 723 and 7 047 new radio AGN candidates in HETDEX and S82 respectively.

Acknowledgments This work has been supported by the Fundação para a Ciência e a Tecnologia (FCT) through the Fellowship PD/BD/150455/2019 (PhD::SPACE Doctoral Network PD/00040/2012) and POCH/FSE (EC) and through research grants PTDC/FIS-AST/29245/2017, UID/FIS/04434/2019, UIDB/04434/2020 and UIDP/04434/2020.

References

1. Ahumada, R., Prieto, C.A., et al.: The 16th data release of the sloan digital sky surveys: first release from the APOGEE-2 southern survey and full release of eBOSS spectra. Astrophys. J. Suppl. **249**(1), 3 (2020). https://doi.org/10.3847/1538-4365/ab929e
2. Carvajal, R., Matute, I., et al.: Exploring new redshift indicators for radio-powerful AGN. Galaxies **9**(4), 86 (2021). https://doi.org/10.3390/galaxies9040086
3. Cutri, R.M., Skrutskie, M.F., et al.: VizieR Online Data Catalog: 2MASS All-Sky Catalog of Point Sources (Cutri+ 2003). VizieR Online Data Catalog, II/246 (2003)
4. Cutri, R.M., Wright, E.L., et al.: VizieR Online Data Catalog: AllWISE Data Release (Cutri+ 2013). VizieR Online Data Catalog, II/328 (2013)
5. de Gasperin, F., Williams, W.L., et al.: The LOFAR LBA sky survey. I. Survey description and preliminary data release. Astron. Astrophys. **648**, A104 (2021). https://doi.org/10.1051/0004-6361/202140316
6. della Ceca, R., Lamorani, G., et al.: The properties of X-ray-selected active galactic nuclei. III. The radio-quiet versus radio-loud samples. Astrophys. J. **430**, 533 (1994). https://doi.org/10.1086/174428
7. Flesch, E.W.: The million quasars (Milliquas) v7.2 catalogue, now with VLASS associations. The inclusion of SDSS-DR16Q quasars is detailed (2021). e-prints, arXiv:2105.12985
8. Flewelling, H.A., Magnier, E.A., et al.: The Pan-STARRS1 database and data products. Astrophys. J. S **251**(1), 7 (2020). https://doi.org/10.3847/1538-4365/abb82d
9. Gordon, Y.A., Boyce, M.M., et al.: A catalog of very large array sky survey epoch 1 quick look components, sources, and host identifications. Res. Note. Am. Astron. Soc. **4**(10), 175 (2020). https://doi.org/10.3847/2515-5172/abbe23
10. Hill, G.J., Gebhardt, K., et al.: The Hobby-Eberly telescope dark energy experiment (HETDEX): description and early pilot survey results. In: Kodama, T., Yamada, T., Aoki, K. (eds.) Panoramic Views of Galaxy Formation and Evolution. Astronomical Society of the Pacific Conference Series, vol. 399, p. 115 (2008)
11. Hodge, J.A., Becker, R.H., et al.: High-resolution very large array imaging of sloan digital sky survey stripe 82 at 1.4 GHz. Astrophys. J. **142**(1), 3 (2011). https://doi.org/10.1088/0004-6256/142/1/3
12. Intema, H.T., Jagannathan, et al.: The GMRT 150 MHz all-sky radio survey. First alternative data release TGSS ADR1. Astron. Astrophys. **598**, A78 (2017). https://doi.org/10.1051/0004-6361/201628536
13. Macfarlane, C., Best, P.N., et al.: The radio loudness of SDSS quasars from the LOFAR two-metre Sky Survey: ubiquitous jet activity and constraints on star formation. Mon. Not. R. Astron. Soc. **506**(4), 5888–5907 (2021). https://doi.org/10.1093/mnras/stab1998
14. Marocco, F., Eisenhardt, P.R.M., et al.: The CatWISE2020 Catalog. Astrophys. J. Suppl. **253**(1), 8 (2021). https://doi.org/10.3847/1538-4365/abd805
15. Norris, R.P., Hopkins, A.M., et al.: EMU: evolutionary map of the universe. Publ. Astron. Soc. Aust. **28**(3), 215–248 (2011). https://doi.org/10.1071/AS11021

16. Padovani, P.: The radio-loud fraction and its dependence on magnitude and redshift. Mon. Not. R. Astron. Soc. **263**, 461–470 (1993). https://doi.org/10.1093/mnras/263.2.461
17. Shimwell, T.W., Tasse, C., et al.: The LOFAR two-metre sky survey. II. first data release. Astron. Astrophys. **622**, A1 (2019). https://doi.org/10.1051/0004-6361/201833559
18. Skrutskie, M.F., Cutri, R.M., et al.: The two micron all sky survey (2MASS). Astrophys. J. **131**(2), 1163–1183 (2006). https://doi.org/10.1086/498708
19. Wright, E.L., Eisenhardt, P.R.M., et al.: The wide-field infrared survey explorer (WISE): mission description and initial on-orbit performance. Astrophys. J. **140**(6), 1868–1881 (2010). https://doi.org/10.1088/0004-6256/140/6/1868

A Machine Learning Suite to Halo-Galaxy Connection

Natalí S. M. de Santi, Natália V. N. Rodrigues, Antonio D. Montero-Dorta, L. Raul Abramo, Beatriz Tucci, and M. Celeste Artale

Abstract As far as we know, galaxies form inside dark matter halos and elucidating this connection is a key element in theories of galaxy formation and evolution. In this work we propose a suite of machine learning tools to predict baryonic from halo properties in the IllustrisTNG300 magnetohydrodynamical simulation. We apply four methods: extremely randomized trees, K-nearest neighbors, light gradient boosting machine, and neural networks. Moreover, we combine the results of them in a stacked model. In addition, we apply all these methods in an augmented dataset using the synthetic minority over-sampling technique for regression with Gaussian noise, to deal with the problem of imbalanced data sets. Altogether, the ML algorithms are consistent at predicting central galaxy properties from a set of input halo properties such as halo mass, concentration, spin, and halo overdensity. For stellar mass, the Pearson correlation coefficient is 0.98, while for specific star formation rate, color, and size it is between 0.7–0.8. Lastly, the presented analysis adds evidence to previous works indicating that certain galaxy properties cannot be reproduced using halo features alone.

N. S. M. de Santi (✉) · N. V. N. Rodrigues · L. R. Abramo
Instituto de Física, Universidade de São paulo, São Paulo, Brazil
e-mail: abramo@if.usp.br

A. D. Montero-Dorta
Departamento de Física, Universidad Técnica Federico Santa María, Valparaíso, Chile

B. Tucci
Max–Planck–Institut für Astrophysik, Garching, Germany

M. C. Artale
Institut für Astro- und Teilchenphysik, Universität Innsbruck, Innsbruck, Austria

Department of Physics and Astronomy, Purdue University, West Lafayette, IN, USA

Physics and Astronomy Department Galileo Galilei, University of Padova, Padova, Italy

F. Bufano et al. (eds.), *Machine Learning for Astrophysics*, Astrophysics and Space Science Proceedings 60, https://doi.org/10.1007/978-3-031-34167-0_7

1 Introduction

The relations between the baryonic and dark matter components of the Universe is one of the most important areas of research in cosmology and galaxy evolution. Different methodologies are used to characterize this connection, and the hydrodynamical simulations are the state-of-art in the subject. However, they have a huge computational cost. In this way, we propose the development of a machine learning (ML) suite, which uses one of these simulations (the IllustrisTNG) to provide a quick and direct map between the dark matter halo and the central galaxy properties.

2 Methodology

Data we use data from the IllustrisTNG magnetohydrodynamical cosmological simulation [1]. We analyze TNG300 box, of 205 h^{-1}Mpc of side length and that adopts the standard ΛCDM cosmology [2]. The input halo properties are: halo mass, age, spin, concentration and overdensity (δ_3, defined as the number density of subhalos within a sphere of radius $R = 3h^{-1}$Mpc). While the central galaxy properties that we are predicting are stellar mass, specific star formation rate (sSFR), radius and color (g-i, derived from the magnitudes).

Model we solve a regression problem to find the map $F(\mathbf{x} \rightarrow y)$ between the halo properties $\mathbf{x}$ and each galaxy feature y. We have used four well known different ML algorithms: extremely randomized trees (ERT) [3], light gradient boosting machine (LGBM) [4], k-nearest neighbours (kNN) [5] and neural networks (NN) [6]. Instead of choosing one single model, and to improve upon them, we combined their predictions in a *stacked model*. It forms linear combinations of different predictors [7] and summarizes their results. Moreover, we used the synthetic minority oversampling technique for regression with Gaussian noise (SMOGN), to deal with the problem of imbalanced data sets obtaining predictions of underrepresented regions of the data [8]. The scores of the predictions were measured using the Pearson correlation coefficient (PCC), which is better as close to one the value is.

3 Results

The main results of the whole analysis comprehends the galaxy-property predictions in terms of the comparison between the stacked models trained using the "raw" (not augmented) and the SMOGN (augmented) data sets and are presented in Fig. 1. First, we show the comparison of the true and the predicted stellar mass distribution (histogram), on the upper left. Second, we analyze the results of the relations between galaxy and halo properties, presenting the galaxy radius versus halo mass relation for the raw (middle upper) and SMOGN (upper right) results. The

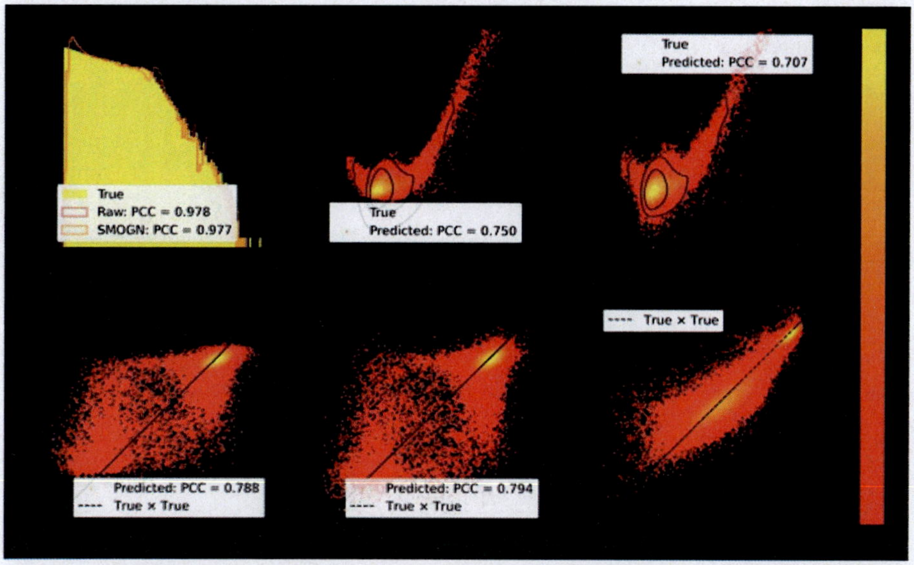

Fig. 1 The upper row shows: a histogram for stellar mass (on the left, comparing the true, raw and SMOGN results), in the middle and in the right we have the scatter representing respectively the raw and SMOGN radius-halo mass relation and the contour represent the true relation. The bottom row shows: the one-to-one basis comparison in scatter for sSFR for the raw (on the left) and SMOGN (on the right). The plot on the bottom right represent this same relation for color, but only for NN predictions, using the raw dataset and including the other galaxy properties as input

scatter plots show the predictions for the normalized density (indicated by the color bar) and the contours represent the same idea, but for true values. The one to one comparison is made for sSFR, in the bottom plots: left, for the stacked model trained with the raw dataset and middle, for stacked model trained with SMOGN dataset. The bottom right represents the same comparison for color, using the predictions for neural networks, while trained with the raw dataset, including as input not only the halo properties but all the other galaxy features. We present the PCC value in all the plots.

4 Discussion and Conclusions

In this work we were able provide combined models to obtain central galaxy properties and to show that the overall distributions are improved using the SMOGN augmented dataset, predicting galaxies in their tails. By construction, ML models tend to concentrate on the bulk of the distributions, which hinders the prediction of scatter even in the metric analysis. SMOGN works in the right direction, increasing the scatter in the relations but slightly increasing the difference between the true and predicted values for the points removed from the peak of the distribution which are taken to the lower and higher ends of it. Thus, the PCC scores for the raw models are still better than the SMOGN results in almost all the cases. A possible solution to recover the complete shape of the predictions, while improving all the scores, can be made using methods to predict the probability distributions, instead of a single value for each object, which will be done in a future work. Finally, the inclusion of other galaxy properties led to improvements in the metrics and a improvement in the one-to-one galaxy properties.

References

1. Pillepich, A., et al.: Simulating galaxy formation with the IllustrisTNG model. Mon. Not. R. Astron. Soc. **473**, 4077 (2018)
2. Planck Collaboration et al.: Planck 2015 results. XIII. Cosmological parameters. Astron. Astrophys. **594**, A13 (2016)
3. Geurts, P.: Extremely randomized trees. Mach. Learn. **63**, 3 (2006)
4. Ke G., Meng Q., Finley T., et al.: Lightgbm: a highly efficient gradient boosting decision tree. Adv. Neural Inf. Process. Syst. **30** (2017)
5. Ivezic, Z., et al.: Statistics, Data Mining, and Machine Learning in Astronomy: A Practical Python Guide for the Analysis of Survey Data. Princeton University Press, Princeton (2014)
6. Chollet, F.: Deep Learning with Python. Manning Publications Company, New York (2017)
7. Breiman L.: Bagging predictors. Mach. Learn. **24** (1996)
8. Branco P., et al.: SMOGN: a Pre-processing approach for imbalanced regression. Proc. Mach. Learn. Res. **74** (2017)

New Applications of Graph Neural Networks in Cosmology

Farida Farsian, Federico Marulli, Lauro Moscardini, and Carlo Giocoli

Abstract Upcoming cosmological surveys will provide unprecedented amount of data, which will require innovative statistical methods to maximize the scientific exploitation. Standard cosmological analyses based on abundances, two-point and higher-order statistics of cosmic tracers have been widely used to investigate the properties of the *cosmic web* and Large Scale Structure. However, these statistics can only exploit a subset of the entire information content available. Our goal is thus to implement new data analysis techniques based on machine learning to extract cosmological information through forward modelling, by directly exploiting the spatial coordinates and other observed properties of galaxies and galaxy clusters. Specifically, we investigated a new representation of large-scale structure data in the form of *graphs*. This data format can be directly fed to *Graph Neural Networks*, a recently proposed class of supervised Deep Learning algorithms. We tested the method on dark matter halo catalogues in different cosmologies, finding promising results. In particular, the method can discriminate among different dark energy models with high accuracy, through both binary classification (99%-accuracy) and multi-class classification (97%-accuracy). Moreover, it provides constraints on the value of w_0, through regression, with high precision.

F. Farsian (✉)
Dipartimento di Fisica e Astronomia "Augusto Righi" - Alma Mater Studiorum, Università di Bologna, Bologna, Italy

Osservatorio di Astrofisica e Scienza dello Spazio di Bologna, Bologna, Italy

Osservatorio Astrofisico di Catania, INAF, Catania, Italy
e-mail: farida.farsian@unibo.it; farida.farsian@inaf.it

F. Marulli · L. Moscardini · C. Giocoli
Bologna University, INAF, INFN, Sezione di Bologna, Bologna, Italy
e-mail: federico.marulli3@unibo.it; lauro.moscardini@unibo.it; carlo.giocoli@inaf.it

F. Bufano et al. (eds.), *Machine Learning for Astrophysics*, Astrophysics and Space Science Proceedings 60, https://doi.org/10.1007/978-3-031-34167-0_8

1 Introduction

Accurate and fast statistical data analysis techniques are essential to extract the full cosmological information content in the data provided by current and next-generation cosmological experiments (e.g. Euclid [1] and LSST). A possible response to these needs could be pulling the information out of raw data available in galaxy or galaxy cluster catalogues. To do so, we propose to consider the large-scale structure data in the form of graphs and let a so-called Graph Neural Network (GNN) to obtain all the specifications of the tracer distribution field.

GNNs are a class of Deep Learning methods designed to perform inferences on irregular and sparse data described by graphs [2]. Indeed, they can capture the graph structure of data which is often very rich, and are particularly suitable for learning global permutation invariant quantities, which makes them ideal to be used with large-scale dark matter and galaxy catalogues [3].

2 Graph Neural Networks

To build a GNN for cosmological inference analyses, we have used the Spektral package,[1] which is based on `Tensorflow` and `Keras`. Our GNN architecture consists of two blocks of EdgeGNN [4] and two blocks of GeneralConv [5], followed by the pooling layers—TopKpool layers [6] and the Global average pooling layer.

We test our GNN model on a large set of dark matter halo catalogues extracted from the Quijote N-body simulations [7], which provide enough statistics to train Neural Networks on cosmologies with different dark energy equation of state parameters. The box size of this simulation is 1 Gpc/h with more than 8.5 trillions of particles at a single redshift. In particular, we consider 500 mock catalogues extracted from the Quijote simulations at redshift $z = 0$, and for three w_0 values: $[-0.95, -1.0, -1.05]$. Massive halos are the most relevant probes for the analysis performed in this work, and we have applied a mass halo cut of $7 \times 10^{14}\,\mathrm{M}_{\odot}$ which means we have around one thousand halos per realization.

The first step of the implemented method consists in building the graph assigned to each dark matter halo catalogue, to give it as input to the GNN. Each halo is considered as a node of the graph, while the halo mass and coordinates are held as node features. Then we assign an edge between nodes. Two nodes, i and j, have an edge ($e_{i \to j}$) if they are closer than a certain distance r. We keep r as one of the hyper-parameters of the method, and implemented a grid search algorithm to optimise its value. The results presented here are obtained with $r = 100\,\mathrm{Mpc/h}$.

[1] https://graphneural.network.

3 Results

To assess the performance of the implemented GNN on the built cosmic graphs, we check its ability in distinguishing and classifying halo graphs with different values of w_0. Then we move to the regression problem, which is more desirable for the purpose of this project. For all the cases, we performed a graph-level analysis on 360 realizations as the training, 40 as the validation and 100 as the test set.

Firstly, we perform a binary classification on $w_0 = [-0.95, -1.05]$. The implemented GNN is able to distinguish among these values with 100% of accuracy in both the training and validation phases, and 99% accuracy for the test set. The next stage of our analysis is dedicated to a multi-class classification, with the three values of w_0 as classes. In this case, the GNN reaches 99% of accuracy in the training and validation phases, and 97% accuracy on the test set.

As a final step, we exploit the GNN to provide a prediction on the value of w_0 for the constructed halo graphs. As it is shown in Fig. 1, the GNN predicts the value of w_0 very accurately, with 2% of statistical error for all the three sets of halo catalogues considered. The error bars show the standard deviation of the GNN predictions for the 100 test set realizations.

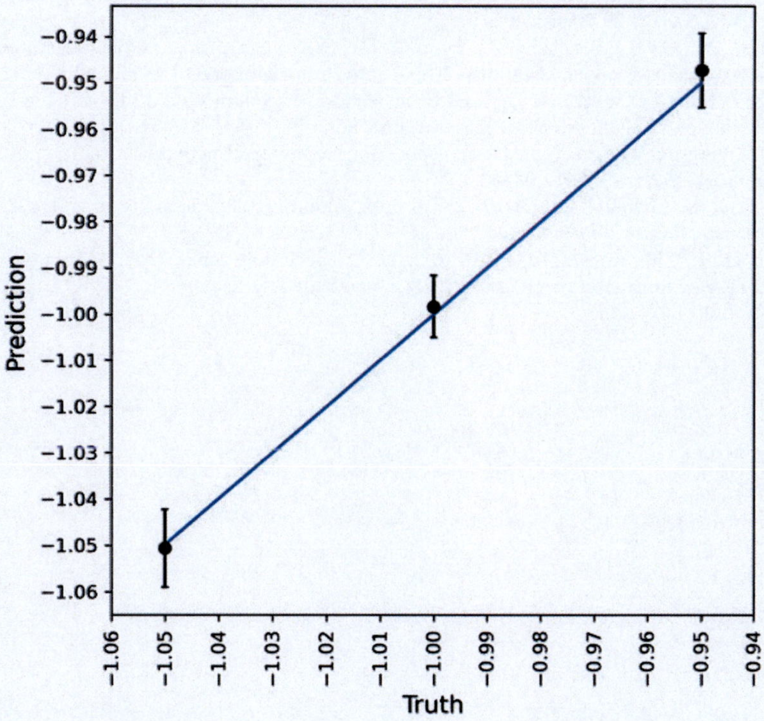

Fig. 1 Predictions of the implemented GNN on w_0. The true w_0 value is shown on the y axis. The error bar indicates the dispersion of GNN prediction among the 100 test set realizations

4 Conclusion and Future Prospects

This work shows that an appropriate GNN can be applied to large-scale structure data sets to capture encoded information, without exploiting summary statistics. The implemented model presented in this work is able to distinguish among dark energy models with different equation of state parameters, with very high accuracy in both binary and multi-class classifications. In the future, we plan to use mock simulations, with different w_0 and w_a values, to validate our GNN on an extended parameter space. After this step, we aim to apply this algorithm to real data sets, to provide constraints on modified gravity models and compare them to the ones from standard statistical analyses.

References

1. Blanchard, A., Euclid Collaboration, et al.: Euclid preparation: VII. Forecast validation for Euclid cosmological probes. Astron. Astrophys. **642**, A191 (2020). https://doi.org/10.1051/0004-6361/202038071.
2. Wu, Z., Pan, S., Chen, F., Long, G., Zhang, C., Yu, P.S.: A comprehensive survey on graph neural networks. IEEE Trans. Neural Netw. Learn. Syst. **32**, 4–24 (2021). https://doi.org/10.1109/TNNLS.2020.2978386.
3. Zhou, J., Cui, G., Hu, S., Zhang, Z., Yang, C., Liu, Z., Wang, L., et al.: Graph neural networks: a review of methods and applications (2018). arxiv, https://doi.org/10.48550/arXiv.1812.08434
4. Wang, Y., Sun, Y., Liu, Z., Sarma, S.E., Bronstein, M.M., Solomon, J.M.: Dynamic graph CNN for learning on point clouds (2018). arxiv https://doi.org/10.48550/arXiv.1801.07829
5. You, J., Ying, R., Leskovec, J.: Design space for graph neural networks (2020). arxiv, https://doi.org/10.48550/arXiv.2011.08843
6. Gao, H., Ji, S.: Graph U-nets (2019). arXiv, https://doi.org/10.48550/arXiv.1905.05178
7. Villaescusa-Navarro, F., Hahn, C., Massara, E., Banerjee, A., Delgado, A.M., Ramanah, D.K., Charnock, T., Giusarma, E., Li, Y., Allys, E., et al.: The quijote simulations. Astrophys. J. Suppl. **250**, 2 (2020). https://doi.org/10.3847/1538-4365/ab9d82

Detection of Point Sources in Maps of the Temperature Anisotropies of the Cosmic Microwave Background

Patricia Diego-Palazuelos, R. Belén Barreiro, Patricio Vielva, David Balbás, Marcos López-Caniego, Diego Herranz, and Biuse Casaponsa

Abstract We address the blind detection of extragalactic point sources on maps of the temperature anisotropies of the cosmic microwave background as an image segmentation problem. We project the sky into two-dimensional patches and design a convolutional neural network (CNN) to identify the sources' position through supervised training with binary labels. To improve the CNN performance, we divide the sky into three regions of progressively increasing Galactic foreground intensity and independently train specialized models for them. With this strategy, we achieve promising completeness levels even in the center of the Galactic plane.

The cosmic microwave background (CMB) is a powerful probe that contains information about inflation, the energy content of the Universe, and the initial perturbations that gave rise to the large-scale structure observed today [1]. To extract this information from microwave observations, we must first separate CMB radiation from the rest of Galactic and extragalactic emissions. In that sense, extragalactic sources emitting in the microwave range, like radio-loud active galactic nuclei or dusty galaxies, constitute a contaminant that one must remove to measure the smallest angular scales of the CMB.

P. Diego-Palazuelos (✉) · R. B. Barreiro · P. Vielva · D. Herranz · B. Casaponsa
Instituto de Física de Cantabria, Santander, Spain
e-mail: diegop@ifca.unican.es

D. Balbás
IMDEA Software Institute, Madrid, Spain

M. López-Caniego
Aurora Technology for the European Space Agency, European Space Astronomy Centre, Madrid, Spain

F. Bufano et al. (eds.), *Machine Learning for Astrophysics*, Astrophysics and Space Science Proceedings 60, https://doi.org/10.1007/978-3-031-34167-0_9

As an alternative to traditional methods, we address the blind detection of extragalactic point sources on maps of the temperature anisotropies of the CMB from a machine learning standpoint. We approach source detection as an image segmentation problem and design a convolutional neural network (CNN) that successfully solves it while maintaining a simple architecture. Here we demonstrate the methodology's potential through its application to realistic simulations.

To train and test the performance of our CNN, we build realistic simulations of the microwave sky as seen by the 143 GHz band of the *Planck* satellite [1]. These mock observations contain CMB realizations according to *Planck*'s best-fit cosmological model [2], realistic simulations of *Planck*'s instrumental noise [1], and Galactic and extragalactic foregrounds. We use the *Planck Sky Model* [3] to simulate maps of the synchrotron, thermal dust, spinning dust, and free-free diffuse Galactic emissions, and the extragalactic thermal and kinetic Sunyaev-Zeldovich effects. We generate extragalactic point sources following the number counts per flux density models of Ref. [4]. We distinguish between detectable (labeled) sources and a background of faint undetectable (unlabeled) sources with fluxes below 177 mJy. This threshold is based on the properties of the Second *Planck* Catalog of Compact Sources [5].

We project the sky into $3.63° \times 3.63°$ patches of 128×128 pixels. Based on their mean Galactic foreground intensity, we divide patches into training sets of faint (60% of the sky), medium (50%), and bright (35%) foreground emission corresponding to the regions of the sky shown in Fig. 1. To compensate its reduced

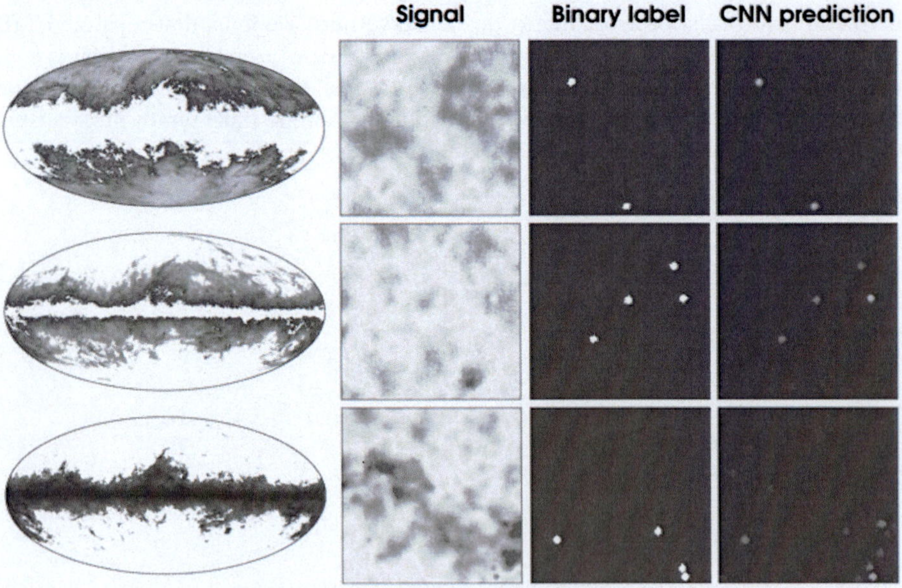

Fig. 1 From top to bottom, division of the sky into regions of faint, medium, and bright Galactic foreground emission, and CNN performance in an example patch from each region

Table 1 CNN architecture. k shows the number of kernels per layer and GN indicates where group normalization [7] was applied before activation. Convolutional layers are evenly padded with zeros. The dropout layer is only included in the CNN trained on the region of brightest Galactic foreground emission

Layer	Input	k	Kernel size	Stride	GN	Activation	Output
Conv.	128 × 128	16	7 × 7	1 × 1	✓	ReLU	128 × 128
Spatial dropout [8]	20% of feature maps are dropped out						
Conv.	128 × 128	32	7 × 7	1 × 1	✓*	ReLU	128 × 128
Max pooling	128 × 128	–	2 × 2	2 × 2	–	ReLU	64 × 64
Transpose conv.	64 × 64	64	5 × 5	2 × 2	✓	ReLU	128 × 128
Conv.	128 × 128	1	1 × 1	1 × 1	–	Sigmoid	128 × 128

* Only when dropout is not applied

sky coverage, we increase the size of the bright training set by adding rotated and re-scaled patches from the medium region that do not overlap with the bright (an extra $\sim$ 20% of the sky). We use supervised learning with binary labels to train independent instances of the autoencoder in Table 1 that specialize in each region. Figure 1 shows an example of the CNN performance in each region. The CNNs trained on the faint and medium regions are well-behaved, while the CNN trained on the bright one shows signs of over-fitting. Preliminary results indicate that the generalization ability of the CNN trained in the bright region improves with regularization techniques.

Once over-fitting is under control, we will be ready to extend the training to other frequencies and, eventually, apply the CNN to actual data. Preliminary results show that, at 143 GHz, our CNN achieves a 90% completeness at 270 mJy in the region of brightest Galactic foreground emission while yielding only a 10% of spurious detections. In addition, compared to traditional methods like the matched filter [6], it is able to reduce the flux at which a 90% completeness is reached from 514 mJy to 195 mJy in the region of intermediate foreground intensity (with 1% of spurious detections allowed). These results lead us to believe that, once applied to data, our CNN will produce deeper and more complete catalogs of extragalactic sources.

Acknowledgments PDP, RBB, PV, and DH acknowledge funding from the Unidad de Excelencia María de Maeztu (MDM-2017-0765). PDP, RBB, and PV thank the Spanish Agencia Estatal de Investigación (AEI, MICIU) for the financial support provided under the project PID2019-110610RB-C21, and DH for that under the project PGC2018-101814-B-I00. PDP also acknowledges funding from the *Formación del Profesorado Universitario* program of the Spanish Ministerio de Ciencia, Innovación y Universidades. We acknowledge the use of the *Planck Sky Model*, `Tensorflow`, `Keras`, `HEALPix`, `Scipy`, `skimage`, `Matplotlib`, and `NumPy`.

References

1. Planck Collaboration: *Planck* 2018 results. I. Overview and the cosmological legacy of *Planck*. Astron. Astrophys. **641**, A1 (2020). https://doi.org/10.1051/0004-6361/201833880
2. Planck Collaboration: *Planck* 2018 results. VI. Cosmological parameters. Astron. Astrophys. **641**, A6 (2020). https://doi.org/10.1051/0004-6361/201833910
3. Delabrouille, J., et al.: The pre-launch *Planck* sky model: a model of sky emission at submillimetre to centimetre wavelengths. Astron. Astrophys. **553**, A96 (2013). https://doi.org/10.1051/0004-6361/201220019
4. Tucci, M., et al.: High-frequency predictions for number counts and spectral properties of extragalactic radio sources. New evidence of a break at mm wavelengths in spectra of bright blazar sources. Astron. Astrophys. **533**, A57 (2011). https://doi.org/10.1051/0004-6361/201116972
5. Planck Collaboration: *Planck* 2015 results. XXVI. The Second *Planck* catalogue of compact sources. Astron. Astrophys. **594**, A26 (2016). https://doi.org/10.1051/0004-6361/201526914
6. Herranz, D., Vielva, P.: Cosmic microwave background images. IEEE Signal Process. Mag. **27**(1), 67 (2010). https://ieeexplore.ieee.org/document/5355497
7. Wu, Y., He, K.: Group normalization. Int. J. Comput. Vis. **128**, 742 (2020). https://doi.org/10.1007/s11263-019-01198-w
8. Tompson, J., et al.: Efficient object localization using convolutional networks. In: 2015 CVPR Conference, vol. 648 (2015). https://ieeexplore.ieee.org/document/7298664

Reconstruction and Particle Identification with CYGNO Experiment

A. Prajapati and E. Baracchini

Abstract We investigate the potential of using deep learning techniques to identify possible background events in search of Dark Matter with directional time projection chamber. The difference in the topological features can be learned by the deep learning models by training over thousand of events. These networks can be further used for the discrimination of background and signal events. The networks trained in this study performs better than the conventional approach (applying selection on variable) used for the classification of backgrounds and signal events. There is potential to further improve the performance of these models to improve the background rejection capabilities.

1 The CYGNO Experiment

The goal of the CYGNO project is to boost the advancement of gaseous Time Projection Chambers (TPC) detectors for directional DM searches and solar neutrino spectroscopy. Weakly interactive massive DM Particles (WIMPs) can interact with ordinary matter nuclei, inducing Nuclear Recoils (NR)[1]. Assuming the Standard galactic Halo Model, the energy of the induced NR is expected to be of the order

---

CYGNO Collaboration (F. D. Amaro, L. Benussi, S. Bianco, C. Capoccia, M. Caponero, D. S. Cardoso, G. Cavoto, A. Cortez, I. A. Costa, G. D'Imperio, E. Dané, G. Dho, F. Di Giambattista, E. Di Marco, F. Iacoangeli, H. P. Lima Junior, G. S. P. Lopes, G. Maccarrone, R. D. P. Mano , R. R. Marcelo Gregorio, D. J. G. Marques, G. Mazzitelli, A. G. McLean, A. Messina, C. M. B. Monteiro, R. A. Nobrega, I. F. Pains, E. Paoletti, L. Passamonti, S. Pelosi, F. Petrucci, S. Piacentini, D. Piccolo, D. Pierluigi, D. Pinci, F. Renga, R. J. D. C. Roque, F. Rosatelli, A. Russo, G. Saviano, N. J. C. Spooner, R. Tesauro, S. Tomassini, S. Torelli, and J. M. F. dos Santos)

A. Prajapati (✉) · E. Baracchini
Gran Sasso Science Institute, L'Aquila, Italy

Istituto Nazionale di Fisica Nucleare (INFN)- Laboratori Nazionali del Gran Sasso, L'Aquila, Italy
e-mail: atul.prajapati@gssi.it

F. Bufano et al. (eds.), *Machine Learning for Astrophysics*, Astrophysics and Space Science Proceedings 60, https://doi.org/10.1007/978-3-031-34167-0_10

of few keVs, sufficient to produce a detectable track inside a TPC. The track inside the TPC produces ionization and a stack of 3 Gas Electron Multipliers (GEMs) is used to amplify the ionization produced during the passage of the particle and as consequence visible light is emitted by the CF_4 scintillator. The scintillation photons are detected by the sCMOS camera and by a set of fast photo-sensors, namely Photo-Multiplier Tubes (PMTs). The high granularity and very low noise camera acquires images on the plane perpendicular to the electron drift direction (X-Y co-ordinate), while the PMTs allow for the reconstruction of Z co-ordinate.

Cygno initial R&D phase started in 2015 and was devoted to the assessment of the gas mixture and to find the best readout strategy. Several prototypes were built, from ORANGE[2] (with an active volume of 0.1 L) to the latest LIME prototype[3] (50 L). Cygno is now in the so called PHASE-0: LIME prototype, currently operating underground at the Laboratori Nazionali del Gran Sasso (LNGS), paving the way for the so called PHASE-1, aiming the construction of a $0.4\,m^3$ detector, as upgrade of the LIME prototype. This detector will be commissioned at LNGS in 2023/2024. The final aim of the collaboration is to build a larger detector of 30–$100\,m^3$.

2 Simulation and Discriminating Variables

One of the major challenges is identifying background electron recoil (ER) events caused by gamma rays from radioactive contaminants in the detector materials and surrounding environment. To simulate background ER and expected signal we use GEANT4 and SRIM respectively[4]. Simulated sCMOS images are obtained by adding the detector effects to the simulated track (like effective ionisation, diffusion, amplification, gain fluctuation, light production and geometrical acceptance) and then adding the camera noise recorded with sCMOS camera. Fully digitised images are reconstructed for energy and other physical variables with an iterative density based scanning algorithm (IDBSCAN)[5], which is done in 3 steps as follows:

1. The single pixel noise suppression
2. Clustering: Using IDBSCAN to find three types of clusters based on the energy density and spatial density of the track
3. Superclustering: Tracks are created by merging two or more clusters found in step 2.

Interaction of different particles with nuclei or the electrons in the gas of the TPC produces different patterns in the gas and thus in the 2D projection of the tracks. Track shape observables can hence be used to distinguish NR and ER. In order to exploit these, we compute the following variables[6]: integral, track length, standard deviation of charge distribution, cylindrical thickness, maximum density and Length along principal axis, light density, slimness etc.

3 Models and Training

The track shape variables are used to train the deep learning models. There are 5000 background and 3000 signal events in a total of 12 energy bins in the range of 2–36 keV, of which 80% of the events are used for training and 20% for testing. Energy bins are as follows: 2,4,6,8,10,12,14,16,22,26,30 and 36 keV (Fig. 1).

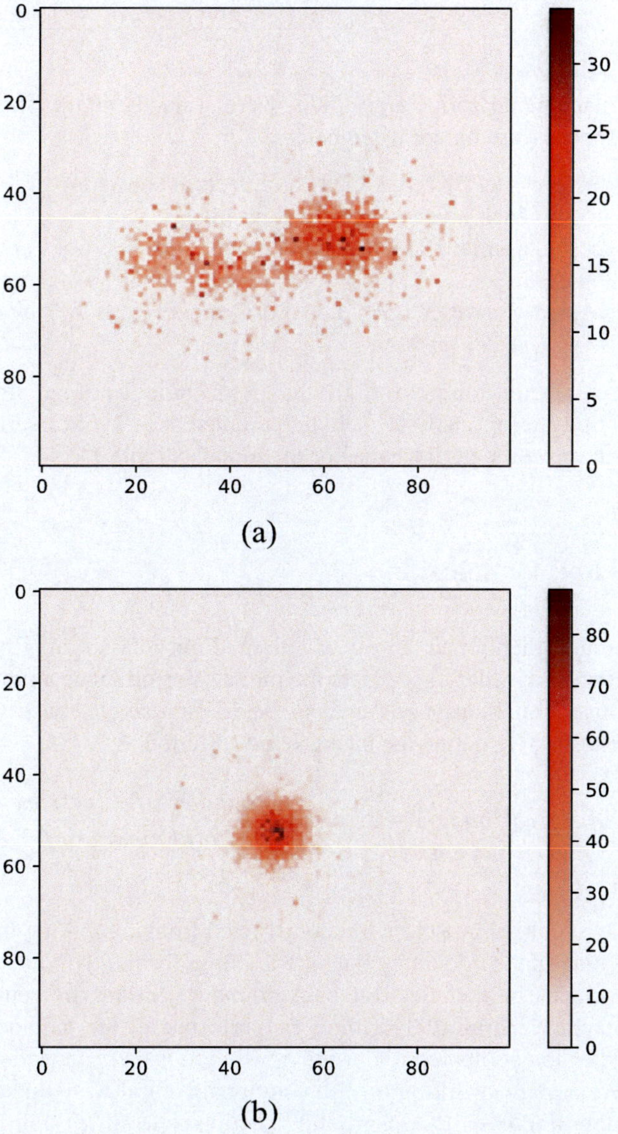

Fig. 1 30 keV electron recoil and nuclear recoil tracks simulated and reconstructed with the method discussed in Sect. 2. (**a**) 30 keV ER track. (**b**) 30 keV NR track

Table 1 Table shows the background rejection for all the 3 machine learning algorithm and rejection obtained with cut-based approach discussed earlier. Rejection values are quoted for a signal efficiency of 40% and 50%

Models	Signal Eff. [ϵ^{S}]%	Bkg. Rej. [1-ϵ^{B}]%
RFC	40	99.54
	50	98.78
GBC	40	99.38
	50	98.55
DNN	40	99.43
	50	98.50
Cut-based	40	83.13
	50	67.20

Different machine learning algorithms were trained on the shape variables. Models used for the training are listed below:

- Deep Neural Network (DNN): A DNN architecture consisting of 3 hidden layers and 10 neurons in each layer is used for this study.
- Random Forest Classifier (RFC): A RFC model with 300 trees and a maximum depth of 13 is used.
- Gradient Boosted Classifier (GBC): A GBC model with 300 trees is used for training.

A fine hyperparameter tuning for all the 3 machine learning algorithms was performed to find the optimal set of hyperparameters to avoid overfitting on this dataset and maximize the performance of the models (Table 1).

4 Results and Conclusion

All the 3 models mentioned above are trained for classifying the signal and background events. In order to compare the models performance among themselves and with a conventional cut-based analysis based on a single shape variable (light density i.e. delta)[4], we define the background rejection as,

$$Background\ Rejection = 1 - \epsilon^{\mathrm{B}} = 1 - \frac{Number\ of\ Electron\ Recoil\ left}{Total\ number\ of\ ER\ events} \tag{1}$$

Comparing each multivariate classifier constructed from combining the track shape variables, we find that introducing machine learning for multivariate analysis leads to an improvement in the electron background rejection. In general, we find that all the machine learning algorithms outperforms all the methods of electron background rejection compared to our previously published conventional cut-based analysis[4]. We are also working on Convolutional Neural Networks (CNN) based models to further improve the background rejection capabilities (Fig. 2).

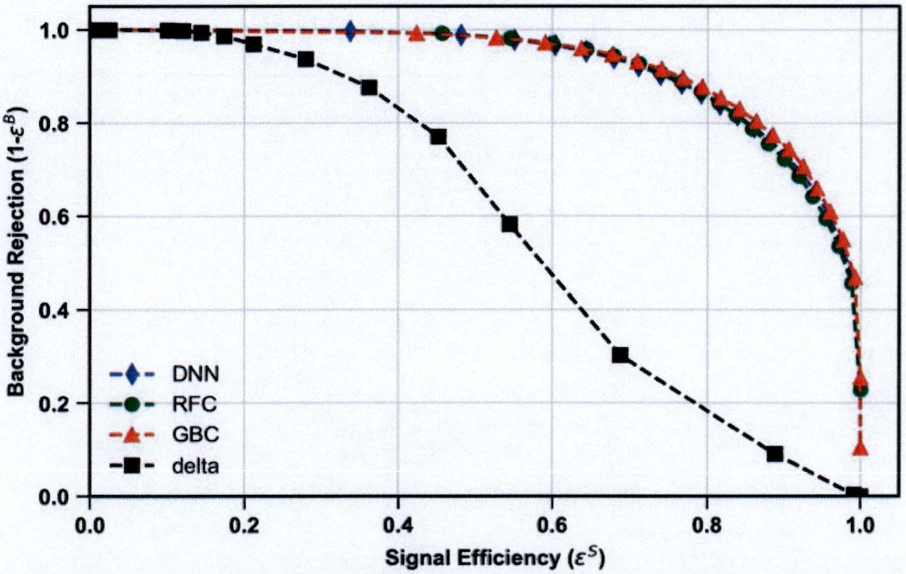

Fig. 2 Figure shows the background rejection as a function of signal efficiency. All the 3 machine learning algorithm show high background rejection compared to rejection obtained using classical approach shown by black curve. Black curve shows the background rejection obtained by applying cuts on the variable light density (delta)

Acknowledgments This project has received fundings under the European Union's Horizon 2020 research and innovation programme from the European Research Council (ERC) grant agreement No 818744.

References

1. Bertone, G., Hooper, D., Silk, J., Particle dark matter: evidence, candidates and constraints. Phys. Rep. **405**(5–6), 279–390 (2005). https://doi.org/10.1016/j.physrep.2004.08.031
2. Marafini, M., et al.: ORANGE: a high sensitivity particle tracker based on optically read out GEM. Nucl. Instrum. Meth. A **845**, 285–288 (2017). https://doi.org/10.1016/j.nima.2016.04.014
3. Amaro, F. D., Baracchini, E., et al.: The CYGNO experiment, MPDI (2022). https://doi.org/10.3390/instruments6010006
4. Baracchini, E., et al.: Identification of low energy nuclear recoils in a gas time projection chamber with optical readout. Meas. Sci. Technol **32**, 025902 (2020)
5. Baracchini, E., et al.: A density-based clustering algorithm for the CYGNO data analysis. J. Instrum. **15.** (2020). https://doi.org/10.1088/1748-0221/15/12/T12003
6. Ghrear, M., Vahsen, S.E. & Deaconu, C.: Observables for recoil identification in high-definition gas time projection chambers. J. Cosmol. Astropart. Phys. (2020) https://doi.org/10.1088/1475-7516/2021/10/005

Event Reconstruction for Neutrino Telescopes

Philipp Eller

Abstract Neutrino telescopes, such as IceCube or KM3NeT, consist of thousands of photo sensors distributed over cubic kilometer volumes. The Cherenkov light from particle interactions within creates signals of varying shapes and sparsity. For the reconstruction of such physics events, one aims to infer quantities like the interaction vertex, the deposited energy, the angles, and the topology of the interaction. This reconstruction step is of central relevance to many physics analyses and searches, and improvements in both accuracy and speed have a direct, positive impact on the science. This proceeding will shortly review some common reconstruction methods, and present a few novel event reconstruction algorithms based on machine learning.

1 Neutrino Telescopes

Optical deep-water or deep-ice neutrino telescopes are a unique type of instrument that allow the detection of neutrinos in the GeV–PeV energy range. Observed neutrinos are produced predominantly in the Earth's atmosphere, but neutrinos of astrophysical origin have also been detected [1, 2], opening the field of neutrino astronomy. Multiple such telescopes have been constructed, including ANTARES [3] and IceCube [4, 5], both successfully collecting over 10 years of data each, with the latter still in operation for the foreseeable future. Multiple new projects are under construction (KM3NeT [6], Baikal-GVD [7], and IceCube-Upgrade [8, 9]) or are proposed (IceCube Gen2 [10], P-ONE [11], or TRIDENT [12]). All of these detectors have in common that they are built from so-called "strings" or "lines" submerged in a natural optical medium (water or ice), carrying a number of modules. These modules encapsulate one or several photo-multiplier tubes (PMTs)

P. Eller (✉)
Department of Physics, TU Munich, Munich, Germany
e-mail: philipp.eller@tum.de

F. Bufano et al. (eds.), *Machine Learning for Astrophysics*, Astrophysics and Space Science Proceedings 60, https://doi.org/10.1007/978-3-031-34167-0_11

in a pressure housing. Multiple strings are arranged in a, generally, irregular pattern to make up a full telescope.

Neutrinos interacting inside or close to the detector volume can create a detectable trace of Cherenkov radiation that is emitted from charged particles produced in the interaction. Such events can result in only a handful of modules detecting light (this is the case in general for low energy interactions, or interactions far from strings), or "light up" large parts of the detector creating stunning signatures spanning many hundreds of meters for very high energy events. Figure 1 shows an example event display for a generic neutrino telescope.

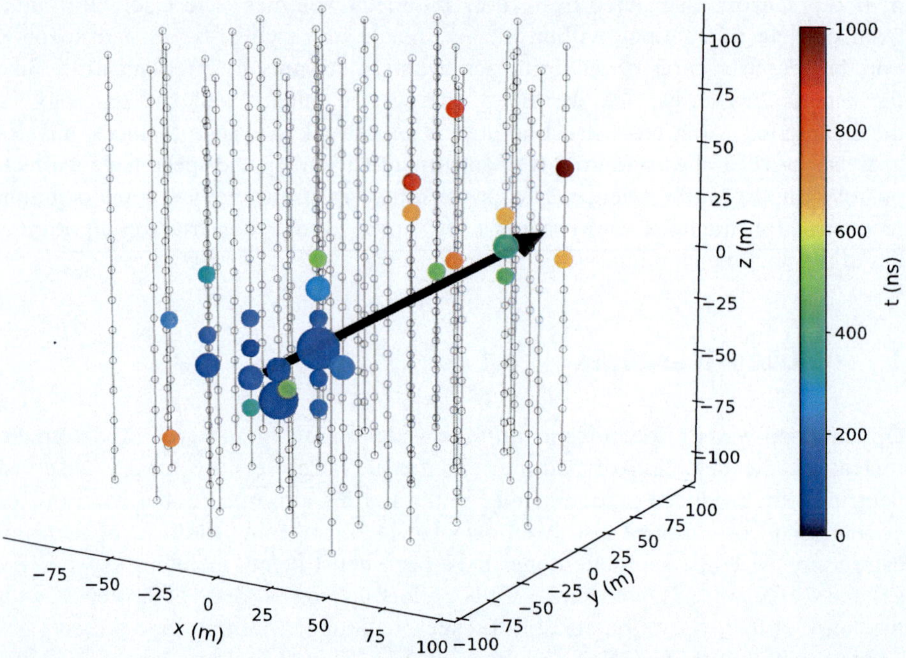

Fig. 1 Example event display of a neutrino interaction in a generic neutrino telescope. The light gray circles connected via vertical lines display optical sensors in water or ice (so-called "strings" or "lines"). The black arrow indicates the true position and direction of the interacted neutrino, here a 150 GeV interaction with an outgoing muon track carrying 20% of the energy, while the rest of the energy is dumped into the nucleus resulting in a shower at the vertex point. The colorful dots represent the light observed at the photodetectors. The size indicates the amount of light registered and the color shows the time of the earliest photon

2 Event Reconstruction

Event reconstruction aims at translating the (raw) detector data x—the registered light in each module in an event—to physically interpretable quantities θ, such as an event's energy, interaction vertex, or direction (pointing). This processing step is necessary for almost all scientific analyses of neutrino telescope data. For instance, a search for astrophysical point-sources of neutrino emission needs accurate and precise estimates of the direction neutrinos came from as well as their energies.

In the following, we will review some of the strategies taken for reconstruction of neutrino telescope events, and also discuss some recent approaches based on machine learning (ML).

2.1 Traditional Approaches

Some of the simplest means of event reconstruction can be summarized under the term of **least-squares fitting**, which do not aim to model the detailed detector response but rather directly fit a geometric model to the observed hit pattern. These can be simple straight lines to approximate extended muon tracks in the detector, the center-of-gravity of hits to estimate the interaction vertex, or involve slightly more complex parabola geometry to account for the Cherenkov emission angle [13, 14].

To make use of as much information in the data as possible, however, more sophisticated approaches are needed to model the full detector response including the neutrino interaction, the optical transport (e.g. absorption, scattering) in the detector medium, PMT simulation, or the electronic readout. Based on a response model, a likelihood function $\mathcal{L}(x, \theta)$ can be formulated to correlate the event reconstruction parameters θ with the observed hit pattern in the detector x. With a likelihood at hand, one can use standard inference techniques, such as maximum likelihood or Bayesian inference, to obtain reconstruction estimates and uncertainties.

Different **approximate likelihoods** have been introduced, including those of ANTARES [15], IceCube [16–20], or IceCube-DeepCore [14]. These usually rely on tabulated detector response functions obtained via simulation and other parameterizations and approximations to make the likelihood evaluations fast and deterministic. A way to largely avoid resorting to approximations is to base the likelihood directly on forward event simulation (Monte Carlo), which usually is available to a high degree of fidelity. Repeated simulation can be used to directly sample the expectations. Such approaches have been used [21], but are computationally extremely costly and can in practice only be used for few events.

2.2 ML Regression

In contrast to a likelihood based method tying together observations x and parameters θ, that then can be used to find desirable estimators of event parameters $\hat{\theta}$, ML regressors work in a different way. The latter will directly return an estimator given the data $x \rightarrow \hat{\theta}$, or often given transformed data $\tilde{x}$. These ML models usually come with a large number of weights that need to be adjusted during a training process, and the solution depends on the specific choice of a loss function.

While dense neural networks, so called multi-layer perceptrons (MLPs), are the canonical architecture to apply, those work only well on heavily transformed (or summarized) data that comes at the price of large information loss and hence degraded performance. To input data closer to its raw form, dense NNs start to become too large in terms of free weights if a detector features thousands of channels.[1]

Instead, inspired by image processing, correlating only neighboring data together can be beneficial. In practice, the so-called **Convolutional Neural Networks** (CNNs) [22] take input data on a regular pixel grid and process them by applying several layers of filters (shared weights). Such approaches have been demonstrated to work for IceCube [23], or KM3NeT ORCA [24]. One major downside is the need to transform the data into a form of a regular pixel grid. In general (and in all practical terms), detector arrays are irregular, and data appear in variable-length sequences. The transformation onto fixed length, rigid image-like structure inevitably results in some information loss. One way to reduce the need to summarize detector hits and to keep the important time structure is the use of **Recurrent Neural Networks** (RNNs) [22], that can also be mixed with CNNs as recurrent sub-blocks. Such approaches are under active development at neutrino telescopes. The method of **Graph Neural Networks** [25] is a generalization of CNNs from pixel grids onto mathematical graphs. Not only can features (nodes) be specified, but also their interconnection (edges). This means that, for instance, the pair-wise distance of sensors can be encoded naturally to allow for arbitrary detector geometries. Such GNNs have recently been employed for reconstruction and classification in IceCube [26, 27].

The aforementioned ML methods can be trained to predict point estimates for single or multiple parameters at once, and simple error estimates can also be predicted.

[1] For example, the IceCube detector features 5160 channels. A simple MLP with one hidden layer of size 256 and a single output would amount to 1,321,473 trainable weights.

Table 1 Overview over different reconstruction strategies used in neutrino telescopes

Method	Strengths	Weaknesses
Least-squares fitting	Simple, robust, fast	Inaccurate
Approximate likelihood	Physics driven, interpretable, general purpose, accurate	Often slow, approximations necessary
CNN/RNN	Established ML architectures, fast evaluation	Lossy data transformations necessary, inflexible, training loss definition
GNN	Natural representation of arbitrary detector geometries	Relatively complex ML models, training loss definition
Generator	More interpretable than regressors	Slower than regressors for point estimates
Ratio estimator	Flexible (ragged inputs), minimal assumptions	Slower than regressors for point estimates

2.3 *Hybrid ML-Likelihood Approaches*

This third category of methods can be understood as a hybrid of the first two. For the **Generator** approach, similar NN architectures as discussed before can be employed, but instead of predicting an estimator given the data, the direction is reversed, and the NN predicts the (distribution of) expected observations given a set of event parameters $\theta \rightarrow x$. Apart from using such an approach for the fast generation of random data samples, it can also serve as basis to define a reconstruction likelihood.

Taking this thought one step further, a neat trick in statistics can be exploited to directly learn a function proportional to the likelihood with minimal assumptions $x, \theta \rightarrow \hat{r} \propto \mathcal{L}(x, \theta)$. The "trick" here is to use a correlated training set of $p(\theta, x)$ together with an un-correlated one $p(\theta), p(x)$. A classifier trained on the binary cross-entropy loss aiming at distinguishing samples from the two populations is the **Ratio Estimator** r. An optimal classifier $\hat{r}$ can be shown to approximate the likelihood-to-evidence ratio [28]. Such an approach has been demonstrated to work well for neutrino telescope event reconstruction in Ref. [29].

The following Table 1 provides a **short summary** of the discussed (no-exhaustive) list of reconstruction methods for neutrino telescopes.

3 Conclusions

The data from neutrino telescopes has to undergo a reconstruction process before being used for scientific data analysis. Throughout, different Ansatzes haven been proposed and used in the field, including simple least-square fitting, methods approximating the likelihood, or various techniques based on machine learning. All approaches come with their strength and weaknesses, that were reviewed in this paper. In general, high inference speed is a clear advantage, and improvements in both accuracy and speed have a direct, positive impact on the science.

References

1. Aartsen, M.G., et al.: Science **342**, 1242856 (2013). https://doi.org/10.1126/science.1242856
2. Aartsen, M.G., et al.: Phys. Rev. Lett. **113**, 101101 (2014). https://doi.org/10.1103/PhysRevLett.113.101101
3. Ageron, M., et al.: Nucl. Instrum. Meth. A **656**, 11 (2011). https://doi.org/10.1016/j.nima.2011.06.103
4. Aartsen, M.G., et al.: J. Instrum. **12**(03), P03012 (2017). https://doi.org/10.1088/1748-0221/12/03/P03012
5. Abbasi, R., et al.: Astropart. Phys. **35**(10), 615 (2012)
6. Adrian-Martinez, S., et al.: J. Phys. G **43**(8), 084001 (2016). https://doi.org/10.1088/0954-3899/43/8/084001
7. Avrorin, A.D., et al.: EPJ Web Conf. **191**, 01006 (2018). https://doi.org/10.1051/epjconf/201819101006
8. Aartsen, M.G., et al.: Letter of Intent: The Precision IceCube Next Generation Upgrade (PINGU) (2014)
9. Aartsen, M.G., et al.: J. Phys. G **44**(5), 054006 (2017). https://doi.org/10.1088/1361-6471/44/5/054006
10. Aartsen, M.G., et al.: J. Phys. G **48**(6), 060501 (2021). https://doi.org/10.1088/1361-6471/abbd48
11. Agostini, M., et al.: Nature Astron. **4**(10), 913 (2020). https://doi.org/10.1038/s41550-020-1182-4
12. Ye, Z.P., et al.: Proposal for a neutrino telescope in South China Sea (2022)
13. Aguilar, J.A., et al.: Astropart. Phys. **34**, 652 (2011). https://doi.org/10.1016/j.astropartphys.2011.01.003
14. Abbasi, R., et al.: Eur. Phys. J. C **82**(9), 807 (2022). https://doi.org/10.1140/epjc/s10052-022-10721-2
15. Albert, A., et al.: Astron. J. **154**(6), 275 (2017). https://doi.org/10.3847/1538-3881/aa9709
16. Ahrens, J., et al.: Nucl. Instrum. Meth. A **524**, 169 (2004). https://doi.org/10.1016/j.nima.2004.01.065
17. Aartsen, M.G., et al.: J. Instrum. **9**, P03009 (2014). https://doi.org/10.1088/1748-0221/9/03/P03009
18. Abbasi, R., Abdou, Y., Ackermann, M., Adams, J., Aguilar, J., Ahlers, M., Altmann, D., Andeen, K., Auffenberg, J., Bai, X., et al.: Nucl. Instum. Meth. A **703**, 190–198 (2013)
19. Aartsen, M.G., et al.: Nucl. Instrum. Meth. A **736**, 143 (2014). https://doi.org/10.1016/j.nima.2013.10.074
20. Abbasi, R., et al.: J. Instrum. **16**(08), P08034 (2021). https://doi.org/10.1088/1748-0221/16/08/P08034
21. Chirkin, D.: 33rd International Cosmic Ray Conference (2013)
22. LeCun, Y., Bengio, Y., Hinton, G.: Nature **521**(7553), 436 (2015)
23. Abbasi, R., et al.: J. Instrum. **16**, P07041 (2021). https://doi.org/10.1088/1748-0221/16/07/P07041
24. Aiello, S., et al.: J. Instrum. **15**(10), P10005 (2020). https://doi.org/10.1088/1748-0221/15/10/P10005
25. Scarselli, F., Gori, M., Tsoi, A.C., Hagenbuchner, M., Monfardini, G.: IEEE Trans. Neural Netw. **20**(1), 61 (2009). https://doi.org/10.1109/TNN.2008.2005605

26. Abbasi, R., et al.: J. Instrum. **17**(11), P11003 (2022). https://doi.org/10.1088/1748-0221/17/11/P11003
27. Søgaard, A., Ørsøe, R.F., Bozianu, L., Holm, M., Iversen, K.E., Guggenmos, T., Ha Minh, M., Eller, P., Petersen, T.C.: J. Open Source Softw. **8**(85), 4971 (2023). https://doi.org/10.21105/joss.04971
28. Hermans, J., Begy, V., Louppe, G.: Likelihood-free MCMC with amortized approximate ratio estimators (2020)
29. Eller, P., Fienberg, A., Weldert, J., Wendel, G., Böser, S., Cowen, D.F.: Nucl. Instrum. Meth. A **1048**, 168011 (2023). https://doi.org/10.1016/j.nima.2023.168011

Classification of Evolved Stars with (Unsupervised) Machine Learning

J. Welsh, C. Bordiu, J. Tan, M. Krokos, F. Bufano, and E. Sciacca

Abstract This paper documents our analysis on the effectiveness of unsupervised machine learning algorithms for classification of evolved stars based on multi-wavelength photometric measurements. The foundation is a custom made reference dataset compiled from available stellar catalogues for target sources—AGB, Wolf Rayet, luminous blue variable and red supergiant stars. Our results indicate that applying HDBSCAN to UMAP's feature representation seems to be the most effective approach for this usecase.

1 Introduction

The study of evolved stars is crucial to understanding and quantifying stellar feedback at a Galactic scale. However, a proper identification of such stars is becoming impractical due to the increasing size of the datasets delivered by next-generation all-sky surveys, often requiring spectroscopic follow-up observations for confirmation. In an attempt to speed up the process, automatic approaches have been tested, mostly focusing on supervised machine learning methods [1]. Typically, supervised learning approaches will face two major problems: firstly, they rely on manually labelled datasets, the creation of which is a time-consuming, error-prone task that may introduce biases; and secondly, the resulting datasets can be greatly imbalanced due to the rarity of some types of stars. Aiming at overcoming these challenges, in this work we explored the application of unsupervised machine

---

J. Welsh (✉) · M. Krokos
School of Creative Technologies, University of Portsmouth, Portsmouth, UK
e-mail: jamie.welsh@myport.ac.uk

C. Bordiu · F. Bufano · E. Sciacca
Istituto Nazionale di Astrofisica, Catania, Italy
e-mail: cristobal.bordiu@inaf.it

J. Tan
School of Computing, University of Portsmouth, Portsmouth, UK

F. Bufano et al. (eds.), *Machine Learning for Astrophysics*, Astrophysics and Space Science Proceedings 60, https://doi.org/10.1007/978-3-031-34167-0_12

learning techniques to disentangle different populations of evolved stars, employing exclusively multiwavelength photometric data retrieved from different surveys like Gaia [2], 2MASS [3] and WISE [4].

2 Methodology

Data Preparation

We built a custom dataset of 15,991 known Galactic compact sources belonging to five unique classes of evolved stars, distributed as follows: 8710 Oxygen-rich AGB stars (54.46%); 6135 Carbon-rich AGB stars (38.36%); 778 red supergiants (4.86%); 340 Wolf Rayet stars (2.12%); and 28 Luminous Blue Variables (0.17%). Then, for each source, we used `TOPCAT` to collect optical, near- and far-infrared photometric magnitudes from Gaia EDR3 (G_{BP} = 532 nm, G_{RP} = 797 nm), 2MASS ($J = 1.25\,\mu$m, $H = 1.65\,\mu$m, $K = 2.17\,\mu$m) and WISE ($W1 = 3.4\,\mu$m, $W2 = 4.6\,\mu$m, $W3 = 12\,\mu$m, $W4 = 22\,\mu$m). For the crossmatch, we considered a radius of 5 arcseconds, taking the object closest to the source coordinates in case of multiple matches. To have input features that convey physically meaningful information, we computed a colour base of eight independent colours, namely $G_{BP} - G_{RP}$, $J - H$, $H - K$, $W1 - W2$, $W2 - W3$, $W3 - W4$, $G_{RP} - J$, and $K - W1$.

Unsupervised Learning Setup

We tested the ability of different clustering algorithms to correctly separate the stars into their respective classes using only the colour features. Two different approaches were analysed: (1) straightforward clustering in the original 8D colour space; and (2) clustering in a lower dimensionality space following a feature extraction stage, in which PCA, autoencoders, UMAP and t-SNE were considered. To measure the clustering performance and obtain a fair comparison between models, we relied on the Rand index distance to the ground truth (i.e. the original class of each point). Both centroid-based (KMeans) and density-based (DBSCAN and HDBSCAN) clustering algorithms were compared.

3 Results

Direct clustering on the 8D space showed poor results with both centroid and density based clustering attaining accuracies of $< 70\%$. However, HDBSCAN showed very promising results in terms of separating stars into meaningful groups, most notably classifying a group purely composed of 41 WR stars. However, results such as these could only be achieved by adjusting hyperparameters such that a large proportion of the data became classed as outliers: The same result showed 8019 sources (approximately 50% of the dataset) being classed as outliers. Efforts to make outlier detection more conservative by adjusting HDBSCAN's hyperparameters caused

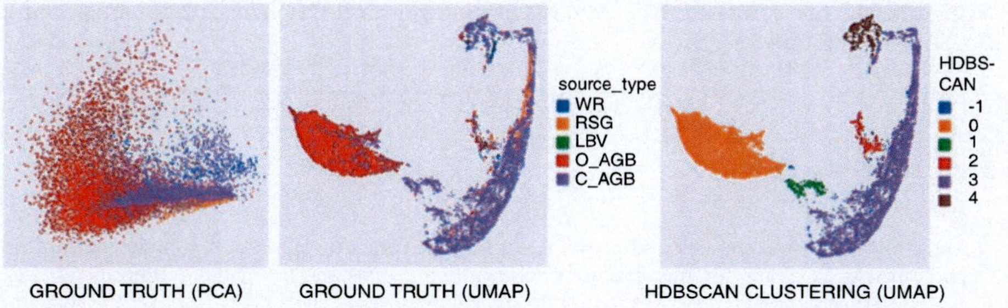

Fig. 1 Visualisation of clustering results: In this result, UMAP is applied to the feature space first, then HDBSCAN is applied to the UMAP feature representation. PCA is used here to demonstrate how the original 8D space might have appeared in 2 dimensions before the application of UMAP

clusters to merge together, rendering the output negligible. When not accounting for outliers, a strong 98.8% accuracy was computed, however the true accuracy computed when considering outliers was 64%.

Adding a dimensionality reduction stage prior to clustering allowed for a more conservative detection of outliers and subsequently improved the clustering accuracy of HDBSCAN. In particular, applying UMAP (n_neighbors = 10, min_dist = 0) before HDBSCAN (min_cluster_size = 12, min_samples = 14) led to 86% accuracy with just 103 outliers. A recently documented approach, (Not Too) Deep Clustering (N2D) [5] which suggests applying UMAP to autoencoded embeddings before the clustering stage, achieved 84% accuracy and zero outliers when using HDBSCAN. Results were particularly impressive with these approaches, producing a clear separation of O-rich AGB (red group) and C-rich AGB (purple group) sources, as shown in Fig. 1, improving upon existing literature results that reported a slight separation in near-infrared colour-colour diagrams due to different dust properties [6, 7].

4 Conclusions

Our results indicate that density-based clustering techniques such as HDBSCAN are superior to centroid-based algorithms when using photometric colours as data input. We found that feature extraction significantly improves clustering accuracy. Methods such as UMAP, that focus on preserving global structure over local structure, perform better, especially when combined with HDBSCAN. The results obtained so far are encouraging, highlighting the scientific value of this unsupervised approach for exploring complex datasets [8]. When applied to unclassified sources, this method may potentially lead to the discovery of new evolved stars in large all-sky surveys.

Acknowledgments The research has received funding from the EC Horizon 2020 grant agreement No. 863448 (NEANIAS).

References

1. Maravelias, G., Bonanos, A.Z., Tramper, F., de Wit, S., Yang, M., Bonfini, P.: A machinelearning photometric classifier for massive stars in nearby galaxies i. The method. Preprint (2022). arXiv:2203.08125
2. Collaboration, G., et al.: The gaia mission (2016). Preprint arXiv:1609.04153
3. Skrutskie, M., Cutri, R., Stiening, R.,Weinberg, M., Schneider, S., Carpenter, J., Beichman, C., Capps, R., Chester, T., Elias, J., et al.: The two micron all sky survey (2mass). Astron. J. **131**(2), 1163 (2006)
4. Wright, E.L., Eisenhardt, P.R., Mainzer, A.K.,Ressler, M.E., Cutri, R.M., Jarrett, T., Kirkpatrick, J.D., Padgett, D., McMillan, R.S., Skrutskie, M., et al.: The wide-field infrared survey explorer (wise): mission description and initial on-orbit performance. Astron. J. **140**(6), 1868 (2010)
5. McConville, R., Santos-Rodriguez, R., Piechocki, R.J., Craddock, I.: N2d: (not too) deep clustering via clustering the local manifold of an autoencoded embedding. In: 2020 25th International Conference on Pattern Recognition (ICPR), pp. 5145–5152. IEEE (2021)
6. Walker, H., Cohen, M.: The classification of stars from iras colors. Astron. J. **95**, 1801–1816 (1988)
7. Van der Veen, W., Habing, H.: The iras two-colour diagram as a tool for studying late stages of stellar evolution. Astron. Astrophys. **194**, 125–134 (1988)
8. Cecconello, T., Bordiu, C., Bufano, F., Puerari, L., Riggi, S., Schisano, E., Sciacca, E., Maruccia, Y., Vizzari, G.: Latent space explorer: unsupervised data pattern discovery on the cloud. Preprint (2022). arXiv:2204.13933

Patterns in the Chaos: An Unsupervised View of Galactic Supernova Remnants

C. Bordiu, F. Bufano, T. Cecconello, E. Sciacca, S. Riggi, and G. Vizzari

Abstract Supernova Remnants (SNRs) constitute a remarkably heterogeneous population that exhibits a wide variety of observational properties. These differences are likely determined by the progenitor star nature, inhomogeneities in the pre-SN circumstellar material, explosion dynamics and other factors difficult to constrain. Is there any underlying order in this apparent chaos? To systematically search for patterns in the Galactic SNR population, we employed an unsupervised approach that, feeding on multiwavelength imagery, allows for an unbiased clustering of SNRs in a multidimensional latent space. Here, we present a pipeline consisting of (i) a feature extraction stage that achieves a compact representation of the sources using infrared and radio continuum imagery; and (ii) a clustering stage that groups together objects that display similar features in the resulting latent space. By applying this pipeline to a representative sample of Galactic SNRs, we found several physically meaningful subsets sharing common observational features.

1 Motivation

Supernova Remnants (hereafter SNRs) are the leftovers of dead stars, the debris that remains after a supernova explosion. These expanding structures made of dust and gas, bounded by shockwaves and carrying large amounts of energy and momentum, are crucial to understand the structural, dynamical and chemical evolution of the Galaxy: they contribute to the chemical enrichment of the surrounding ISM by spreading nucleosynthesis products; and induce star formation by compressing nearby molecular clouds. So far, 383 SNRs have been discovered in the Galaxy [1], constituting a rather heterogeneous population: while most SNRs are radio

C. Bordiu (✉) · F. Bufano · E. Sciacca · S. Riggi
INAF-Osservatorio Astrofisico di Catania, Catania, Italy
e-mail: cristobal.bordiu@inaf.it; filomena.bufano@inaf.it

T. Cecconello · G. Vizzari
University of Milano Biccoca, Milano, Italy
e-mail: thomas.cecconello@unimib.it

F. Bufano et al. (eds.), *Machine Learning for Astrophysics*, Astrophysics and Space Science Proceedings 60, https://doi.org/10.1007/978-3-031-34167-0_13

emitters, owing to thermal radiation from hot plasma and synchrotron, only a fraction of them are visible at mid- and far-infrared wavelengths [2], that typically trace their dust content. From a morphological standpoint, SNRs exhibit a wide range of features, from round shells to intricate filamentary structures. All these observational properties are determined by many factors: the inherent explosion dynamics, inhomogeneities of the pre-SN circumstellar medium, the nature of the progenitor star and its mass-loss history, and many others.

Here we present preliminary results of an experimental study that, using an unsupervised deep learning approach and multiwavelength imagery, aims at finding patterns in a representative sample of the Galactic SNR population.

2 Methodology

We devised an unsupervised pipeline consisting of a representation learning stage based on a convolutional autoencoder, followed by a clustering stage operating on the resulting latent space.

Data Selection and Preparation

We collected available infrared and radio continuum imagery of the SNR population, using WISE (22 μm), HiGAL (70 μm) and archival radio data of the Galactic Plane (30 cm). The choice of these bands allows for a comprehensive picture of the processes at work in these sources, i.e. the joint contribution from dust (warm and hot) and plasma. The final sample, considering the limited coverage of the HiGAL and radio surveys and the removal of sources either too compact or heavily affected by imaging artefacts, consisted of ~180 sources, for which we extracted postage-stamp cutouts in each band. Then, for each source, the three cutouts were aligned to the same projection, convolved to a common angular resolution and re-gridded on a 64×64 pixel grid, composing an RGB image (R=30 cm, G=70 μm, B=22 μm) of size $3 \times 64 \times 64$. Then, compact sources (e.g. stars or background galaxies) not related to the SNRs were masked on a channel-by-channel basis, employing the CAESAR source finder [3]; and a circular region of interest was defined around each SNR. Finally, all the channels were applied a min-max scaling for normalization.

Representation Learning

We built a standard convolutional autoencoder, depicted in Fig. 1, consisting of three encoder layers and three decoder layers with 64:128:256 filters, using Leaky ReLU as the activation function. The latent space (bottleneck) dimensionality was set to 64, as a compromise between the compression factor (~ 200:1), and the quality of the reconstruction. The autoencoder was trained for 30,000 epochs, using the *adam-amsgrad* optimizer, and a custom MSE loss weighted by the masks of the compact sources. To improve generalization and avoid overfitting (due to the limited number of samples), we employed data augmentation, applying rotation and shifting.

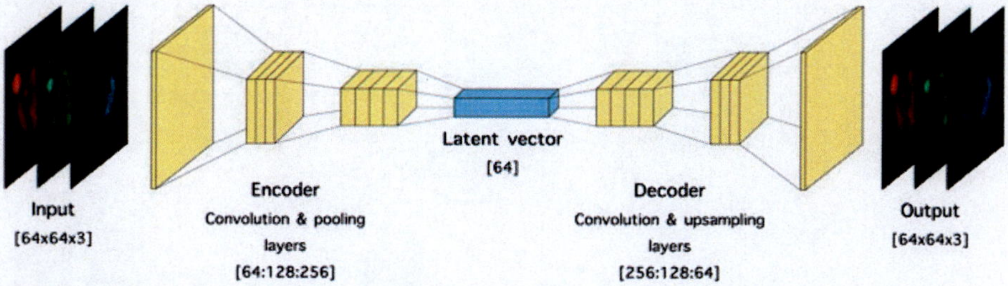

Fig. 1 Sketch of the convolutional autoencoder

Visualization and Clustering

To visualize the results, we employed UMAP, projecting the autoencoded embeddings in a 2D space. We tested different clustering strategies on the 64D latent space. Since this is an exploratory work –we do not know in advance the number of clusters, if any–, we favoured density-based algorithms like DBSCAN.

3 Results

The autoencoder was able to successfully reproduce the input images, while correctly interpolating the masked regions corresponding to the compact sources. It produced a compact representation of the original 180 input images ($3 \times 64 \times 64$ px) in a 64 dimensions latent space. The distribution of points in the latent space was found to be quite sparse, which is not surprising considering the heterogeneous nature of the SNR population and the limited number of samples. For this reason, density-based algorithms performed poorly, with little success at isolating meaningful groups. To overcome this issue, we explored various alternatives, from modifying the CAE architecture to testing other clustering approaches. Finally, inspired by the *Not-too-deep* method described in [4], we employed a UMAP projection of the autoencoded embeddings as an intermediate stage before clustering. We fine-tuned the UMAP hyperparameters, setting `mindist` to 0.0 as recommended by the authors, and exploring a range of *ndim* based on blind guesses of the expected number of clusters (i.e. values in the range 4–8). This approach improved the previous results by a large margin, allowing us to isolate groups of sources that share some common features. In Fig. 2 we present a clustering on a 4D manifold projection of the original 64D embeddings, using DBSCAN ($\epsilon = 0.35$). We find several clusters composed of sources with similar features, both in terms of morphology and colour, e.g. the pink cluster (elongated shells with dominant radio emission), and the green cluster (partial radio shells surrounded by prominent infrared emission).

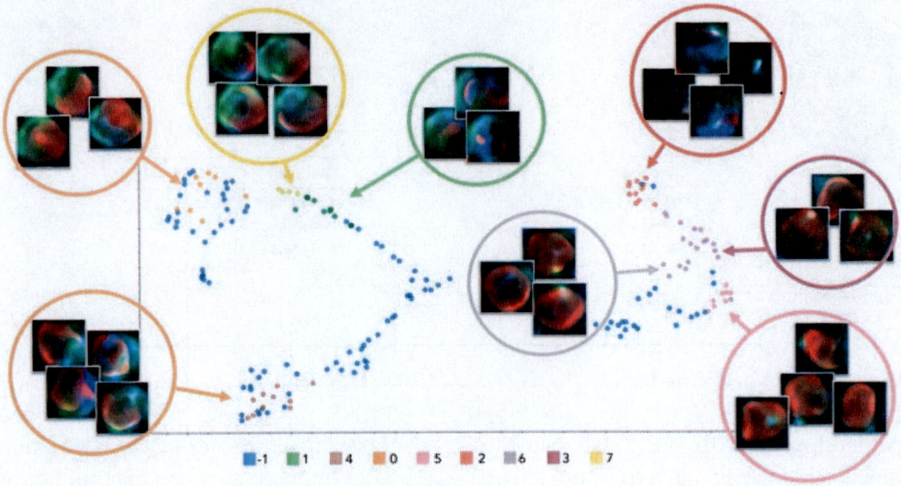

Fig. 2 Clustering of a 4D UMAP manifold projection of the original 64D autoencoded embeddings ($\epsilon = 0.35$). UMAP 2D used for visualization

4 Conclusions

The results obtained so far are promising, highlighting the potential of this approach to perform a systematic study of the Galactic SNR population. The clusters seem to be reliable, as they show little dependence on slight modifications of the architecture hyperparameters. This indicates that, despite the sparse nature of the latent space, which is actually a "continuum" of features, the clusters are physically meaningful. A more in-depth description of this work, including the physical explanation of the clusters (e.g. relation to nature of the progenitor, characteristics of the environment, etc.) will be presented in an upcoming paper [5].

Acknowledgments The research has received funding from the EC Horizon 2020 grant agreement No. 863448 (NEANIAS).

References

1. Ferrand, G., Safi-Harb, S.: A census of high-energy observations of Galactic supernova remnants. Adv. Space Res. **49**(9), 1313–1319 (2012)
2. Chawner, H., Gomez, H.L., Matsuura, M., Smith, M.W.L., Papageorgiou, A., Rho, J., Noriega-Crespo, A., De Looze, I., Barlow, M.J., Cigan, P., Dunne, L., Marsh, K.: A complete catalogue of dusty supernova remnants in the Galactic plane. Month. Not. R. Astron. Soc. **493**(2), 2706–2744 (2020)

3. Riggi, S., Ingallinera, A., Leto, P., Cavallaro, F., Bufano, F., Schillirò, F., Trigilio, C., Umana, G., Buemi, C.S., Norris, R.P.: Automated detection of extended sources in radio maps: progress from the SCORPIO survey. Month. Not. R. Astron. Soc. **460**(2), 1486–1499 (2016) https://arxiv.org/abs/1605.01852arXiv:1605.01852 [astro-ph.IM]. https://doi.org/10.1093/mnras/stw982
4. McConville, R., Santos-Rodríguez, R., Piechocki, R.J., Craddock, I.: N2d: (not too) deep clustering via clustering the local manifold of an autoencoded embedding. In: 2020 25th International Conference on Pattern Recognition (ICPR), pp. 5145–5152 (2021)
5. Bufano, F., et al.: Sifting the debris: patterns in the SNR population with unsupervised ML methods (in preparation)

Clustering of Galaxy Spectra: An Unsupervised Approach with Fisher-EM

J. Dubois, D. Fraix-Burnet, and J. Moultaka

Abstract We present a novel approach to galaxy spectra classification using Fisher-EM, a latent subspace clustering and Gaussian mixture model based algorithm. This approach was applied to a sample of 10,000 simulated spectra, highlighting its capacity to discriminate physical properties based on spectroscopic data, as well as its robustness towards noise. A sample of 700,000 spectra of close-by galaxies observed by the Sloan Digital Sky Survey (SDSS) was successfully classified, and a detailed physical interpretation of the classes is in preparation. An extension to higher redshifts is currently in progress, using a sample of 70,000 galaxies of redshift $0.4 < z < 1.2$ from the VIMOS Public Extragalactic Survey (VIPERS). An evolution tree-like structure was constructed, showcasing evolution pathways of the classes throughout cosmic time from $z = 1.2$ to $z = 0.4$.

1 Introduction

The JWST is now in operation, and the world of astrophysics is about to enter a new era. Observations are made at greater redshifts, and now is a particularly exciting time to study galaxies and their evolution.

The motivation behind this work is to highlight the diversity and evolution of the physical aspects of galaxies that can be inferred from their spectroscopic properties (e.g. metallicity, history of star formation, active nuclei, etc.). In particular, working with a broad range of redshifts makes it possible to evaluate the evolution of these characteristics through the history of the universe, which is ultimately the goal of this work. Using an unsupervised clustering algorithm, an automatic classification of large samples of galaxy spectra is made possible [1]. Galaxies sharing similar

J. Dubois (✉) · D. Fraix-Burnet
Institut de Planétologie et d'Astrophysique de Grenoble (IPAG), Saint-Martin-d'Hères, France
e-mail: julien.dubois@univ.grenoble-alpes.fr; didier.fraix-burnet@univ.grenoble-alpes.fr

J. Moultaka
Institut de Recherche en Astrophysique et Planétologie (IRAP), Toulouse, France
e-mail: jmoultaka@irap.omp.eu

F. Bufano et al. (eds.), *Machine Learning for Astrophysics*, Astrophysics and Space Science Proceedings 60, https://doi.org/10.1007/978-3-031-34167-0_14

properties are gathered in classes, and the physical interpretation can thus be restricted to the classes' mean spectrum instead of the whole sample, hence greatly reducing the computational cost.

The work presented in this contribution consists of an overview of several applications of the unsupervised classification algorithm Fisher-EM to three different datasets of optical spectra. First, on a simulated sample [2], then on observations from the Sloan Digital Sky Survey (SDSS) [3], and finally, on observations from the VIMOS Public Extragalactic Redshift Survey (VIPERS). Another contribution closely related to this work was made by M. Siudek. It showcases Fisher-EM applications to photometric data from VIPERS [4, 5].

The unsupervised algorithm is briefly explained in Sect. 2. An overview of the results that have been obtained is presented in Sects. 3 and 4.

2 Method

The algorithm Fisher-EM was used for this project. It combines both an unsupervised approach based on a Gaussian Mixture Model (GMM), and a dimension reduction process through projection of the data onto a discriminative latent subspace. The description of Fisher-EM is kept concise here, but a complete and thorough explanation can be found in the original article [6].

A GMM models the data's probability density function (PDF) $f(K, \theta)$ with the weighted sum of K multivariate Gaussian PDFs $\Phi(\mu_i, \Sigma_i)$ (Eq. 1). In Fisher-EM, a dimension reduction step is performed before looking to optimize the GMM. The dimension is reduced to $K - 1$ by projecting the data onto a discriminant latent subspace obtained by minimizing the Fisher criterion. The algorithm then aims at minimizing a loss-function by tweaking the number of classes K, the shapes of the GMM's multivariate Gaussian PDFs $\Phi(\mu_i, \Sigma_i)$ and their weight π_i.

$$f(K, \theta) = \sum_{i=1}^{K} \pi_i \Phi(\mu_i, \Sigma_i) \tag{1}$$

3 Physical Relevance and Robustness of the Classification

The physics of galaxies is encoded in their spectra; the presence of emission lines can for example be linked to episodes of star formation and activity in the galactic nuclei; the shape of the continuum gives an indication of the age of the stars; lots of different aspects in a spectrum can be linked back to many physical characteristics. However, a spectrum also contains a lot of non-physical particularities that can influence the classification. For example, the amount of noise is not necessarily identical in every observation; there can be differences in calibration, sampling, and

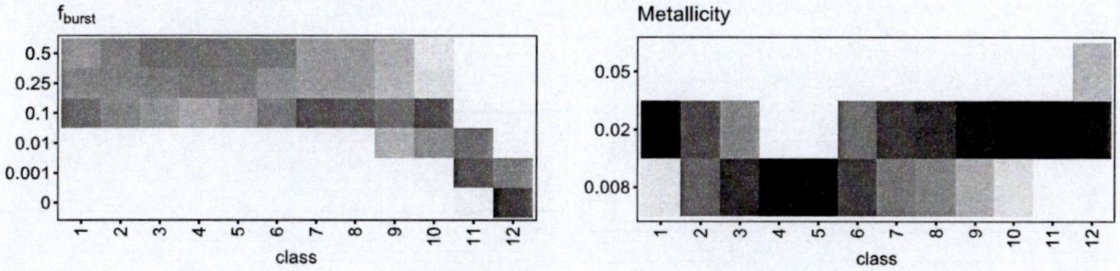

Fig. 1 Heatmaps showing the distribution of the mass fraction of stars originating from a sudden burst of star formation (left panel), and the metallicity (right panel) in the classification of a simulated sample of galaxy spectra. The parameter values are represented on the y-axis, and the class index on the x-axis. The within-class densities of the parameter values are illustrated in the form of a heatmap, where a dark square equates to a density of 1, and white of 0

other instrumental aspects. And because all these components are mixed together in a spectrum, there was a need to test the ability of Fisher-EM to produce classifications that are physically relevant, i.e. that segregate galaxies based on physical characteristics.

Results on optical spectra simulated with CIGALE [7] show that it is in fact the case [2]. Some physical characteristics are very well separated among the classes. The presence and the mass fraction of a burst of star formation, for example, is very well segregated (Fig. 1).

4 A Window to Galaxy Evolution

The method was successfully applied to a sample of 700,000 spectra of close-by galaxies from the SDSS [3]. This is the first automatic classification of a large sample of galaxy spectra performed without any prior feature selection. The resulting classification contains 86 classes, each showing their own specificities, and an in-depth description of the classes is currently in preparation.

An extension of the SDSS classification is currently in progress, and it shall include galaxies from the VIPERS PDR-2 up to a redshift of $z = 1.2$. The sample was divided into subsamples by bins of epoch, and each subsample was individually classified with Fisher-EM. Links between classes of successive epochs were constructed with k-NN to make it possible to follow the evolution of a given class over the cosmic time from 4 Gyr to 9 Gyr after the Big Bang.

When put together, the links create a tree-like structure, where branches can be followed to track evolution pathways (Fig. 2). Preliminary analysis shows a dozen of significant branches with clear spectral specificities, mostly divided into two main categories: star forming and passive. Morphological, spectral, and photometric measurements are being cross-matched with the classes, and will give precise insights for the interpretation of the evolution branches.

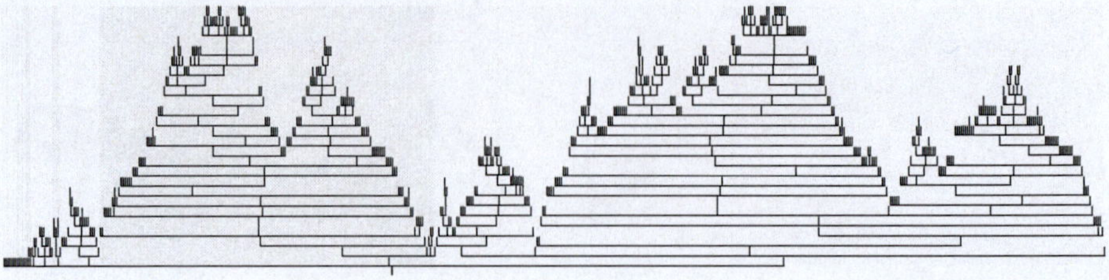

Fig. 2 This tree-like structure highlights evolution pathways of galaxy classes over cosmic time up to a redshift of z = 1.2. Each vertical step in the tree corresponds to a certain epoch, linearly sampled from 4 Gyr after the Big Bang (bottom of the tree) to 9 Gyr (top of the tree)

References

1. Fraix-Burnet, D., Thuillard, M., Chattopadhyay, A.K.: Front. Astron. Space Sci. **2**(3) (2015). https://doi.org/10.3389/fspas.2015.00003. http://www.frontiersin.org/milky_way_and_galaxies/10.3389/fspas.2015.00003/abstract
2. Dubois, J., Fraix-Burnet, D., Moultaka, J., Sharma, P., Burgarella, D.: Astron. Astrophys. **663**, A21 (2022). https://doi.org/10.1051/0004-6361/202141729
3. Fraix-Burnet, D., Bouveyron, C., Moultaka, J.: Astron. Astrophys. **649**, A53 (2021). https://doi.org/10.1051/0004-6361/202040046. https://ui.adsabs.harvard.edu/abs/2021arXiv210305928F
4. Siudek, M., Małek, K., Scodeggio, M., Garilli, B., Pollo, A., Haines, C.P., Fritz, A., Bolzonella, M., de la Torre, S., Granett, B.R., Guzzo, L., Abbas, U., Adami, C., Bottini, D., Cappi, A., Cucciati, O., De Lucia, G., Davidzon, I., Franzetti, P., Iovino, A., Krywult, J., Le Brun, V., Le Fèvre, O., Maccagni, D., Marchetti, A., Marulli, F., Polletta, M., Tasca, L.A.M., Tojeiro, R., Vergani, D., Zanichelli, A., Arnouts, S., Bel, J., Branchini, E., Ilbert, O., Gargiulo, A., Moscardini, L., Takeuchi, T.T., Zamorani, G.: Astron. Astrophys. **597**, A107 (2017). https://doi.org/10.1051/0004-6361/201628951. http://www.aanda.org/10.1051/0004-6361/201628951
5. Siudek, M., Małek, K., Pollo, A., Krakowski, T., Iovino, A., Scodeggio, M., Moutard, T., Zamorani, G., Guzzo, L., Garilli, B., Granett, B.R., Bolzonella, M., de la Torre, S., Abbas, U., Adami, C., Bottini, D., Cappi, A., Cucciati, O., Davidzon, I., Franzetti, P., Fritz, A., Krywult, J., Le Brun, V., Le Fèvre, O., Maccagni, D., Marulli, F., Polletta, M., Tasca, L.A., Tojeiro, R., Vergani, D., Zanichelli, A., Arnouts, S., Bel, J., Branchini, E., Coupon, J., De Lucia, G., Ilbert, O., Haines, C.P., Moscardini, L., Takeuchi, T.T.: Astron. Astrophys. **617**, A70 (2018). https://doi.org/10.1051/0004-6361/201832784. https://www.aanda.org/10.1051/0004-6361/201832784
6. Bouveyron, C., Brunet, C.: Stat. Comput. **22**(1), 301 (2012). https://doi.org/10.1007/s11222-011-9249-9
7. Boquien, M., Burgarella, D., Roehlly, Y., Buat, V., Ciesla, L., Corre, D., Inoue, A.K., Salas, H.: Astron. Astrophys. **622**, A103 (2019). https://doi.org/10.1051/0004-6361/201834156

Unsupervised Classification Reveals New Evolutionary Pathways

M. Siudek, K. Lisiecki, M. Mezcua, K. Małek, A. Pollo, J. Krywult, A. Karska, and M. Junais

Abstract While we already seem to have a general scenario of the evolution of different types of galaxies, a complete and satisfactory understanding of the processes that led to the formation of all the variety of today's galaxy types is still beyond our reach. To solve this problem, we need both large datasets reaching high redshifts and novel methodologies for dealing with them. The VIPERS survey

M. Siudek (✉)
Institute of Space Sciences (ICE, CSIC), Barcelona, Spain

Institut de Física d'Altes Energies (IFAE), The Barcelona Institute of Science and Technology, Barcelona, Spain
e-mail: msiudek@ifae.es

K. Lisiecki
National Centre for Nuclear Research, Warsaw, Poland

Institute of Astronomy, Faculty of Physics, Astronomy and Informatics, Nicolaus Copernicus University, Toruń, Poland

M. Mezcua
Institute of Space Sciences (ICE, CSIC), Barcelona, Spain

Institut d'Estudis Espacials de Catalunya (IEEC), Barcelona, Spain

K. Małek
National Centre for Nuclear Research, Warsaw, Poland

A. Pollo
National Centre for Nuclear Research, Warsaw, Poland

Astronomical Observatory of the Jagiellonian University, Kraków, Poland

J. Krywult
Institute of Physics, Jan Kochanowski University, Kielce, Poland

A. Karska
Institute of Astronomy, Faculty of Physics, Astronomy and Informatics, Nicolaus Copernicus University, Toruń, Poland

Max-Planck-Institut für Radioastronomie, Bonn, Germany

M. Junais
National Centre for Nuclear Research, Warsaw, Poland

F. Bufano et al. (eds.), *Machine Learning for Astrophysics*, Astrophysics and Space Science Proceedings 60, https://doi.org/10.1007/978-3-031-34167-0_15

statistical power, which observed ∼90,000 galaxies at $z > 0.5$, and the application of an unsupervised clustering algorithm allowed us to distinguish 12 galaxy classes. Studies of their environmental dependence indicate that this classification may actually reflect different galaxy evolutionary paths. For instance, a class of the most passive red galaxies gathers galaxies ∼20% smaller than other red galaxies of a similar stellar mass, revealing the first sample of red nuggets at intermediate redshift. On the other end, a class of blue dwarf galaxies is composed mainly of AGN, challenging commonly used mid-infrared AGN selections.

1 Introduction

Nowadays it is essential to apply efficient classification procedures to process large astronomical datasets to understand the universe and its evolution. Modern deep surveys, such as the Dark Energy Spectroscopic Instrument (DESI; [1]) will gather spectra of millions of objects among an even more unprecedented set of over a billion galaxies from imaging surveys (DESI Legacy Imaging Surveys; [2]). Standard approaches to galaxy classification have relied on a manual or citizen scientists' classification, however, such approaches do not scale well with the unprecedented data-growth rate. The solution is the application of ground-breaking machine learning (ML) and deep learning algorithms. Unsupervised (without any training process) ML techniques, which can extract new knowledge and identify so far unknown populations of objects, are of particular interest and have already shown their potential in discovering new galaxy pathways (see [3] for review). In this work, we discuss the potential of the unsupervised ML Fisher Expectation-Maximization (FEM) algorithm [4] in revealing peculiar classes of the galaxy population.

2 Data

The data used in this proceeding comes from the VIMOS Public Extragalactic Redshift Survey (VIPERS; [5]). Observations were performed with the VIsible MultiObject Spectrograph mounted on the Very Large Telescope at ESO. The survey covers ∼23.5 deg^2 split over two W1 and W4 fields of the Canada-France-Hawaii Telescope Legacy Survey Wide. The final data release consists of 86,775 sources limited to $i_{AB} \leq 22.5$ mag targeting galaxies at redshift range $0.4 \lesssim z \lesssim 1.2$. Additionally, the VIPERS survey observed a sample of broad-line AGN (∼1% of the total VIPERS sample) reaching redshift up to $z \sim 4$.

3 Unsupervised Galaxy Classification

VIPERS galaxies with the highest redshift confidence level (> 90%) were separated using the FEM algorithm into 12 clusters corresponding to 3 red, 3 green, and 5 blue galaxy classes and one of the broad-line AGN by [6]. The clustering algorithm works in a 13-dimensional feature space composed of spectroscopic redshift and 12 rest-frame ultraviolet-through-near-infrared magnitudes normalised to the i-band magnitude. Figure 1 shows the projection of the VIPERS sample onto the two principal dimensions of the respective 12-dimensional latent subspace. The FEM classification was also successfully applied on solely photometric datasets (i.e. using photometric redshifts) achieving high accuracy (∼90%) of reproducing spectroscopic classification as well as high efficiency (>90%) in the separation of stars and broad-line AGN [7]. A multi-epoch approach based on the application of the FEM algorithm to VIPERS ($z \sim 0.7$) and SDSS ($z \sim 0.0$) data allowed us to

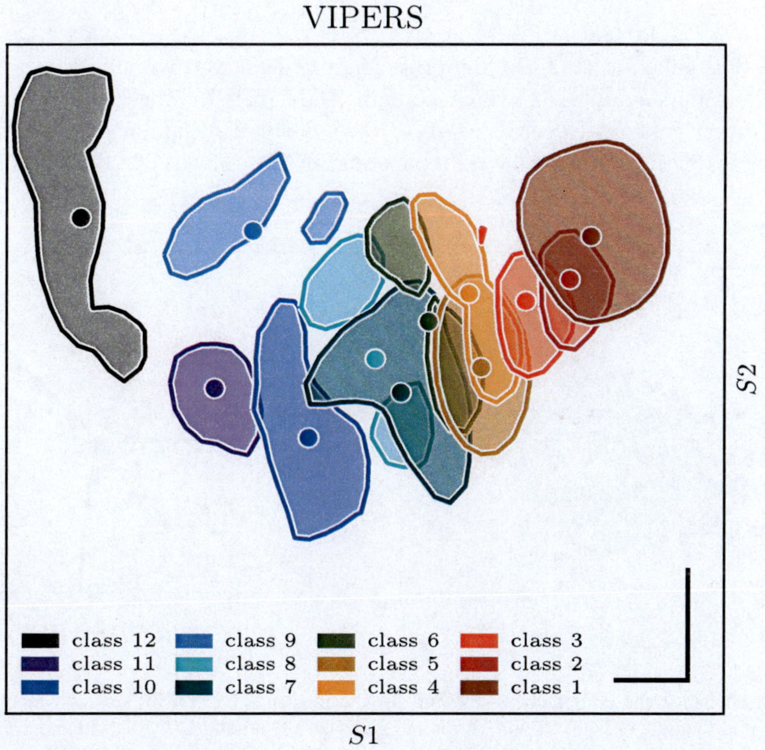

Fig. 1 Projections of VIPERS samples from [6] onto the two dimensions that best separate their clusters. The axes of each plot are determined by FEM and are not corresponding to any peculiar galaxy parameter

study the cosmic evolution of galaxy classes, which we find to mostly reflect the gradual internally-driven growth of bulges and slow quenching [8].

Red Compact Massive Galaxies

The FEM algorithm distinguished 3 red classes, which catch a part of the rare population of red ultra-compact massive galaxies (UCMG) that after formation at high redshift ($z > 2$) have avoided merger processes (red nuggets). In fact, less than 100 UCMG were so far observed at $z < 0.5$ [9] leaving the intermediate redshift ($0.5 < z < 2.0$) unexplored. VIPERS recently filled this niche by providing an unprecedented sample of 77 red nugget candidates [10, 11], of which 72 (94%) are found within FEM red classes. In particular, the majority of them (65%) are found in class 1, containing the most massive and the smallest red galaxies [12]. The red nugget catalogue is created with the most restricted selection cuts (in stellar mass and effective radius, see the left panel in Fig. 2), however, class 1 gathers also larger/less massive galaxies (but still smaller and more massive than standard passive galaxies [13]) further inviting us to explore UCMG in states of their transition.

AGN Dwarf Galaxies

Among blue galaxies, the FEM algorithm identified galaxies with the bluest $NUVr$ and rK colours (class 11), which contain AGN [6, 16]. These are in majority (90%) dwarf galaxies ($\log(M_*/M_\odot) < 9.5$). Their distribution on the emission line diagnostic diagram (see the right panel in Fig. 2) confirms that this class is built

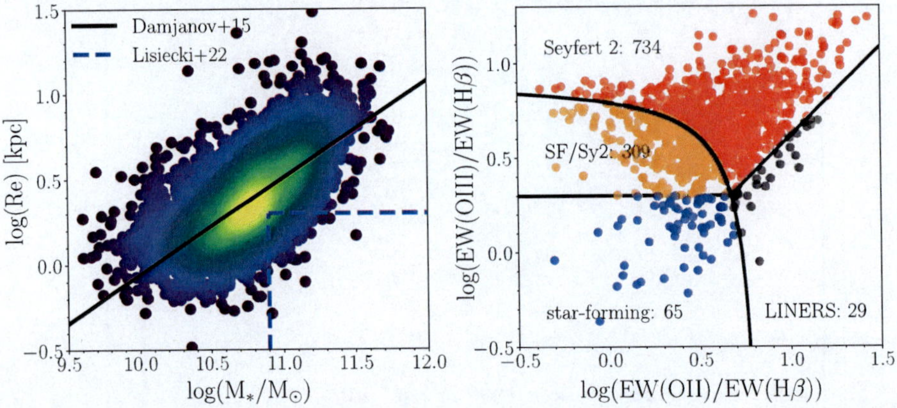

Fig. 2 Left panel: the compactness diagram for the most passive VIPERS class which gathers the first intermediate-redshift sample of red nuggets [10, 11] among UCMG (with a cut proposed by [14]). Right panel: The emission line diagnostic diagram [15] for the bluest VIPERS class which gathers a population of dwarf galaxies hosting AGN [16]

mainly of either AGN (67%) or composite systems (27%[1]) and a small fraction of star-forming galaxies (6%). At the same time, 27% of all spectroscopically-identified VIPERS AGN are gathered in class 11. Such a performance of automatic AGN dwarf selection (accuracy of at least 67% and completeness of 27%) challenges standardly used mid-infrared AGN selection (with an accuracy of ~60% and completeness of ~15%, see Tab. 6 in [17]). Moreover, AGN in dwarf galaxies are nowadays of special interest as most of them are believed to host intermediate-mass black holes, the seeds of supermassive black holes [16, 18].

4 Summary

The unsupervised colour-based VIPERS classification [6] shows great potential for studying physical mechanisms regulating galaxy evolution and automatically revealing classes of peculiar galaxies. In particular, we show that:

- large photometric samples can be used to distinguish different galaxy classes with an accuracy similar to spectroscopic classification [7],
- different evolutionary paths of galaxies are correlated with their environments [12],
- galaxy classes at two different epochs are correlated with each other [8],
- we are able to automatically identify classes of peculiar objects: i) the first intermediate-redshift sample of UCMG [10–12], ii) AGN dwarf galaxies with a performance outshining mid-infrared AGN selections [16].

Acknowledgments This work has been supported by the Polish National Agency for Academic Exchange (Bekker grants BPN/BEK/2021/1/00298/DEC/1 and BPN/BEK/2021/1/00319/DEC/1), the European Union's Horizon 2020 Research and Innovation Programme under the Maria Sklodowska-Curie grant agreement (No. 754510), and the Spanish Ministry of Science and Innovation through the Juan de la Cierva-formacion programme (FJC2018-038792-I), the Polish NCN (UMO-2018/30/E/ST9/00082) and the Ramon y Cajal fellowship (RYC2019-027670-I). This work was also partially supported by the program Unidad de Excelencia María de Maeztu CEX2020-001058-M.

References

1. Abareshi, B., Aguilar, J., et al.: Overview of the instrumentation for the dark energy spectroscopic instrument. Astron. J. **164**, 207 (2022). https://doi.org/10.3847/1538-3881/ac882b
2. Dey, A., Schlegel, D.J., et al.: Overview of the DESI legacy imaging surveys. Astron. J. **157**, 168 (2019). https://doi.org/10.3847/1538-3881/ab089d

[1] It is expected that ~20% of composite objects are AGN. Thus, it is possible that our classification successfully distinguishes AGN and non-AGN galaxies located in the composite region.

3. Baron, D.: Machine learning in astronomy: a practical overview (2019). arXiv:1904.07248
4. Bouveyron, C.,Brunet, C.: Simultaneous model-based clustering and visualization in the Fisher discriminative subspace (2011). arXiv:1101.2374
5. Scodeggio, M., Guzzo, L., et al.: The VIMOS public extragalactic redshift survey (VIPERS). Full spectroscopic data and auxiliary information release (PDR-2). Astron. Astrophys. **609**, A84 (2018). https://doi.org/10.1051/0004-6361/201630114
6. Siudek, M., Małek, K., et al.: The VIMOS public extragalactic redshift survey (VIPERS). The complexity of galaxy populations at 0.4 < z < 1.3 revealed with unsupervised machine-learning algorithms. Astron. Astrophys. **617**, A70 (2018). https://doi.org/10.1051/0004-6361/201832784
7. Siudek, M., Małek, K., et al.: The VIMOS public extragalactic redshift survey (VIPERS). Unsupervised classification with photometric redshifts: a method to accurately classify large galaxy samples without spectroscopic information (2018). arXiv:1805.09905
8. Turner, S., Siudek, M., et al.: Synergies between low- and intermediate-redshift galaxy populations revealed with unsupervised machine learning. Month. Not. R. Astron. Soc. **503**, 3010 (2021). https://doi.org/10.1093/mnras/stab653
9. Scognamiglio, D., Tortora, C., et al.: Building the largest spectroscopic sample of ultracompact massive galaxies with the kilo degree survey. Astrophys. J. **893**, 4 (2020). https://doi.org/10.3847/1538-4357/ab7db3
10. Lisiecki, K., Małek, K., et al.: The first catalogue of spectroscopically confirmed red nuggets at z∼0.7 from the VIPERS survey. Linking high-z red nuggets and local relics. Astron. Astrophys. **669**, A95 (2023). https://doi.org/10.1051/0004-6361/202243616
11. Siudek, M., Lisiecki, K., et al.: Environments of red nuggets at z∼0.7 from the VIPERS survey. Month. Not. R. Astron. Soc. (2023, in press). https://doi.org/10.1093/mnras/stad1685
12. Siudek, M., Małek, K., et al.: Shaping physical properties of galaxy subtypes in the VIPERS survey: environment matters. Astron. Astrophys. **666**, A131 (2022). https://doi.org/10.1051/0004-6361/202243613
13. Siudek, M., Małek, K., et al.: The VIMOS public extragalactic redshift survey (VIPERS). Star formation history of passive red galaxies. Astron. Astrophys. **597**, A107 (2017). https://doi.org/10.1051/0004-6361/201628951
14. Damjanov, I., Zahid, H.J., et al.: The environment of massive quiescent compact galaxies at 0.1 < z < 0.4 in the COSMOS Field. Astrophys. J. **815**, 104 (2015). https://doi.org/10.1088/0004-637X/815/2/104
15. Lamareille, F.: Spectral classification of emission-line galaxies from the sloan digital sky survey. I. An improved classification for high-redshift galaxies. Astron. Astrophys. **509**, A53 (2010). https://doi.org/10.1051/0004-6361/200913168
16. Siudek, M., Mezcua, M.: The environment of AGN dwarf galaxies at z∼0.7 from the VIPERS survey. Month. Not. R. Astron. Soc. **518**, 724 (2023). https://doi.org/10.1093/mnras/stac3092
17. Hviding, R.E., Hainline, K.N., et al.: A new infrared criterion for selecting active galactic nuclei to lower luminosities. Astron. J. **163**, 224 (2022). https://doi.org/10.3847/1538-3881/ac5e33
18. Mezcua, M., Siudek, M., et al.: Astrophys. J. Letters, **943**, L5 (2023). https://doi.org/10.3847/2041-8213/acae25

In Search of the Peculiar: An Unsupervised Approach to Anomaly Detection in the Transient Universe

Dennis Crake and Juan Rafael Martínez-Galarza

Abstract The era of big data time-domain Astronomy is here, with future telescopes increasing the volume of data at an astonishing pace. Perhaps, the most scientifically promising aspect of these surveys is their potential for discovery across the transient universe. Nonetheless, current methods are unable to handle multidimensional databases of such extreme size. Currently, rapid analysis to systematically identify scientifically compelling objects is near impossible. We present an approach that tackles the challenges presented by the onslaught of data, identifying the most anomalous light curves in current and future time-domain surveys. Using observations from the first two years of TESS, we deploy an Unsupervised Random Forest method using a combination of normalized light curve points and their power spectrum as features. Our algorithm identifies a wide range of variability patterns and successfully pinpoints several documented scientifically interesting targets discovered by the community to date.

1 Introduction

Astronomy is undergoing a data-driven revolution, with current telescopes driving the field toward the frontiers of data science and advanced machine learning techniques. The plethora of incoming data brings challenges, dramatically increasing the possibility of overlooking objects with great scientific interest. Future telescopes, such as the Vera C. Rubin observatory, will only exacerbate the issue unless suitable techniques are investigated today.

With past telescopes, the community has relied on serendipitous discoveries by using large teams to trawl through databases in the hope of spotting scientifically

D. Crake (✉)
University of Edinburgh, Royal Observatory Edinburgh, Edinburgh, UK
e-mail: Dennis.Crake@ed.ac.uk

J. R. Martínez-Galarza
Harvard & Smithsonian Centre for Astrophysics, Cambridge, MA, USA
e-mail: juan.martinez.galarza@cfa.harvard.edu

F. Bufano et al. (eds.), *Machine Learning for Astrophysics*, Astrophysics and Space Science Proceedings 60, https://doi.org/10.1007/978-3-031-34167-0_16

interesting objects. Boyajian's star is a great example, undergoing irregular, aperiodic dimming events across multiple days with no sign of an interacting companion [1]. The discovery sparked a great debate about the cause of this dimming pattern within the scientific community and beyond [2–5]. The media attention represents a strong engagement from the general public, justifying the cost of increasingly large and expensive telescopes. As a community, we cannot build the next generation of telescopes and not seize their potential due to a lack of suitable analysis methods. Particularly, as arguably, the most scientifically promising aspect of survey telescopes is the discovery of unknown objects.

This work presents a method of outlier detection to diminish the likelihood of knowledge remaining hidden within these large databases. Nonetheless, to effectively investigate outliers, it is paramount to consolidate the definition of an outlier. For this work, an outlier is an observation that arouses suspicion that it was generated by a different mechanism [6].

2 Data

TESS represents the current generation telescope for exoplanet detection, yet, the data represents an opportunity to explore the greater transient universe. *TESS* can also be used to develop outlier detection techniques in preparation for larger future observatories. This work focuses on the implementation of the Unsupervised Random Forest (URF) technique from [7] (MG21 henceforth) for *TESS* sectors.

The Transiting Exoplanet Survey Satellite (*TESS*) observes the sky in regions known as sectors, each covering a 24×96 degree area. The observations cover approximately 28 days with a 2-minute cadence, producing high-cadence light curves with around 20,000 data points. Each sector also includes a small unobserved period due to maintenance or data transfer. A typical sector release contains 20,000 high-cadence light curves of nearby stars with identical time stamps. Over two years, *TESS* observes 85% of the sky, with a semi-continuous observing region in each hemisphere observed for one year due to the overlap of sectors.

3 The Unsupervised Random Forest

The unsupervised random forest (URF) algorithm is a proven method for outlier detection. An early implementation includes analysis of SDSS galaxy spectra [8] and features in MG21 analysing Kepler light curves. Transforming the random forest algorithm from fundamentally supervised into unsupervised requires an additional synthetic dataset. The training phase focuses on differentiating between the real and synthetic datasets. Therefore, the synthetic dataset must be representative of the original data, whilst being suitably distinguishable by the algorithm. The final dataset is created by a random sample of features from the original dataset,

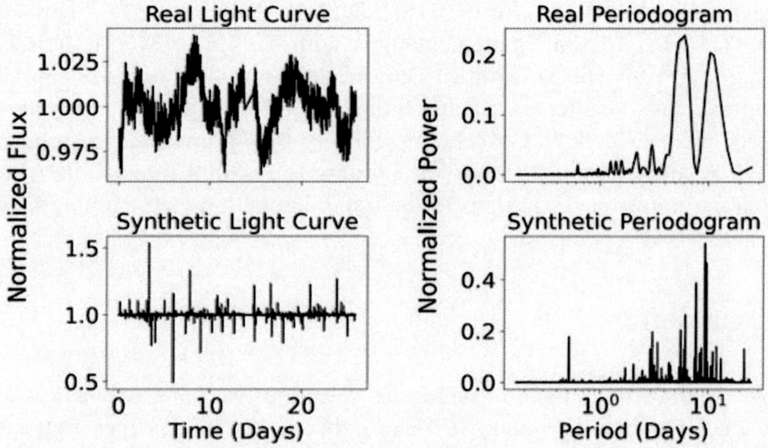

Fig. 1 Example of a real vs synthetic light curve, clearly showing the lack of correlation between features within the synthetic dataset

creating non-physical light curves and periodograms. The resultant dataset mimics the distribution of each feature but breaks the correlations between features. Figure 1 presents an example of a real and synthetic light curve.

The synthetic data is discarded after training, leaving a random forest with numerous nodes populated by the original data. The random forest works by separating dissimilar objects until the real objects occupy separate nodes from the synthetic data. Thus, objects that occupy the same final node are deemed similar. Averaging the node population of a given object over the forest returns the "Anomaly Score", a metric reflecting the isolation of each object. This method of population analysis to calculate the Anomaly Score is statistically equivalent to the pairwise matching approach used in the literature, yet represents an improved complexity from $O(TN^2)$ to $O(TN)$ where $N =$ is the number of objects and $T =$ the number of trees.

3.1 Feature Selection

MG21 discovered it is optimal for the feature selection to remain close to the raw data, which minimises bias during the search for unknown unknowns. Yet, using uncorrected flux data as features is not recommended due to environmental and observer effects. Hence, this work uses normalised flux values with limited spike removal processes with periodogram data to supplement the flux features. The periodogram data probes the variability periodicity beyond flux alone. Unlike Kepler, the light curves are highly populated, resulting in extreme overfitting exacerbated by the smaller population. To avoid overfitting, we investigated several sampling methods for the flux data, with uniform sampling showing the best performance. Bringing knowledge from implementations with Kepler in MG21, it is optimal to use approximately 5000 features. Our analyses revealed a higher emphasis on periodogram data resulted in a substantial outlier population containing $\sim$50% of objects, clear evidence of overfitting. Finally, the optimal outlier population is identified by subsampling the flux to approximately 3000 data points with 1000 periodogram points representing periods from 6 hours to the duration of the sector.

4 Discussion

Through a careful investigation of feature selection, we have successfully implemented a URF to identify outliers in TESS light curves. The unsupervised approach minimises bias on the outliers through the use of a non-physical synthetic dataset for training. The unavoidable limitation of this work is the focus on variations with periods within the duration of a single sector. Future telescopes searching the transient universe do not necessarily observe with a regular cadence meaning future work must aim to remove the current time dependence. A full analysis of results covering 24 sectors is to be presented in an upcoming publication by the authors.

Acknowledgments This work is supported by the UK Science and Technology Facilities Council (STFC) towards the award of a PhD studentship to DC with support from Robert G Mann and Nigel C Hambly. We acknowledge the University of Edinburgh and the Harvard and Smithsonian . | Centre for Astrophysics for their support.

References

1. Boyajian, T.S., LaCourse, D.M., Rappaport, S.A., et al.: Planet hunters IX. KIC 8462852 - where's the flux? Month. Not. R. Astron. Soc. **457**, 3988 (2016). https://doi.org/10.1093/mnras/stw218
2. Meng, H.Y.A., Rieke, G., Dubois, F., et al.: Extinction and the dimming of KIC 8462852. Astrophys. J. **847**, 131 (2017). https://doi.org/10.3847/1538-4357/aa899c
3. Boyajian, T.S., Alonso, R., Ammerman, A., et al.: The first post-kepler brightness dips of KIC 8462852. Astrophys. J. Lett. **853**, L8 (2018). https://doi.org/10.3847/2041-8213/aaa405
4. Foukal, P.: An explanation of the missing flux from Boyajian's mysterious star. Astrophys. J. Lett. **842**, L3 (2017). https://doi.org/10.3847/2041-8213/aa740f
5. Bodman, E.H.L., Quillen, A.: KIC 8462852: transit of a large comet family. Astrophys. J. Lett. **819**, L34 (2016). https://doi.org/10.3847/2041-8205/819/2/L34
6. Hawkins, D.: Identification of Outliers. Springer, Dordrecht (1980)
7. Martínez-Galarza, J.R., Bianco, F.B., Crake, D., et al.: A method for finding anomalous astronomical light curves and their analogues. Month. Not. R. Astron. Soc. **508**, 5734 (2021). https://doi.org/10.1093/mnras/stab2588
8. Baron, D., Poznanski, D.: The weirdest SDSS galaxies: results from an outlier detection algorithm. Month. Not. R. Astron. Soc. **465**, 4530 (2017). https://doi.org/10.1093/mnras/stw3021

Classifying Gamma-Ray Burst X-Ray Afterglows with a Variational Autoencoder

Eliot Ayache and Tanmoy Laskar

Abstract Today, over 1500 gamma-ray burst (GRB) afterglows have been observed by The Neil Gehrels Swift Observatory, with light curves displaying different morphologies in the succession of decay regimes with time. We explore prospects for acquiring physical inference from ML models by investigating the presence of different classes in the morphology of GRB afterglow X-ray light curves. We carry out unsupervised classification of Swift-XRT time-series data using a variational autoencoder. We evaluate our model's ability to identify different morphological classes by carrying out training on synthetic data. The generative aspect of the model can provide physical insight by highlighting the discriminative features in the light curves. We compare the classification results obtained with the traditional functional-form-based classification, and investigate the resulting level of segregation in the data set. We find that the data is segregated in the latent space according to traditional morphology. However, the observed gradual transition in between over-densities unifies the prevalent classification of GRBs based on their X-ray data into a single continuum, suggesting that light curves of different types should be unified under a single model.

1 Introduction

Gamma-ray bursts (GRBs) are produced by collapsing massive stars (long-duration GRBs; LGRBs) or mergers of compact binary systems involving a neutron star (short-duration GRBs; SGRBs). During a GRB event, accretion onto a central

E. Ayache (✉)
The Oskar Klein Centre, Department of Astronomy, Stockholm University, Stockholm, Sweden
e-mail: eliot.ayache@astro.su.se

T. Laskar
Department of Astrophysics/IMAPP, Radboud University, Nijmegen, The Netherlands

Department of Physics & Astronomy, University of Utah, Salt Lake City, UT, USA
e-mail: t.laskar@astro.ru.nl

F. Bufano et al. (eds.), *Machine Learning for Astrophysics*, Astrophysics and Space Science Proceedings 60, https://doi.org/10.1007/978-3-031-34167-0_17

compact remnant (a black hole or a neutron star) leads to the ejection of relativistic, collimated outflows. Internal processes inside these jets produce the prompt γ-rays, which are followed by a long-lived "afterglow" spanning from X-ray to radio wavelengths. In this framework, GRB X-ray afterglows were expected to exhibit simple, power-law light curves (brightness evolution with time). However, the launch of NASA's Neil Gehrels Swift Observatory (*Swift*) in 2005 revealed an unexpectedly rich diversity of X-ray light curves. This discovery challenged the standard afterglow paradigm and spawned a wide variety of theorized emission models (e.g. magnetar winds, disk winds, and jets).

One approach to investigate the underlying emission processes is to perform a classification of GRB X-ray light curves into morphological types. Previous studies [1] attempting such segregation have relied on subjective classification schemes based on the synchrotron model. Here, we present the preliminary results from a study of GRB X-ray light curves with machine learning. We use a variational autoencoder [2] to identify and distil the morphological characteristics of the light curves into a low-dimensional latent space, followed by clustering algorithms applied to the latent space to search for native morphological categories.

2 Data and VAE Model

We select *Swift* X-ray light curves for this study, as this is the single largest, publicly accessible compendium of afterglow light curves for GRBs, which is simultaneously uniform in selection and homogenous in data format. Our data set comprises 1312 X-ray light curves, which we re-bin into 40 temporal bins each, while normalising to preserve the aspect ratio (Fig. 1). In order to test and benchmark our model, we also generate a synthetic data set of realistic X-ray light curves using the parameters of [1] that they fit to a representative subset of the *Swift* data. They classify the light curves according to morphology, by describing the succession of power-law slopes usually observed, into 6 categories (0: single power-law; Ia: one break, steep first; Ib: one break, shallow first; IIa: two breaks, steep, plateau, regular; IIb: two breaks, shallow, steep, shallow; III: oddball). In our analysis, we merge IIb and III together, as they are the least populated categories.

We use a variational autoencoder (VAE) to embed the pre-processed data in a 2D latent space. The generative aspect of the model allows us to better interpret the relationship between samples close together in the latent space by studying the light curves generated by sampling at arbitrary locations. We perform spectral clustering in the latent space (10-nearest-neighbour graph, followed by k-means in the k-dimensional embedded space) to identify clusters. We have experimented with various model architectures, types of layers (including convolutional layers and recurrent layers), and hyper-parameters. Based on tests on a smaller subset

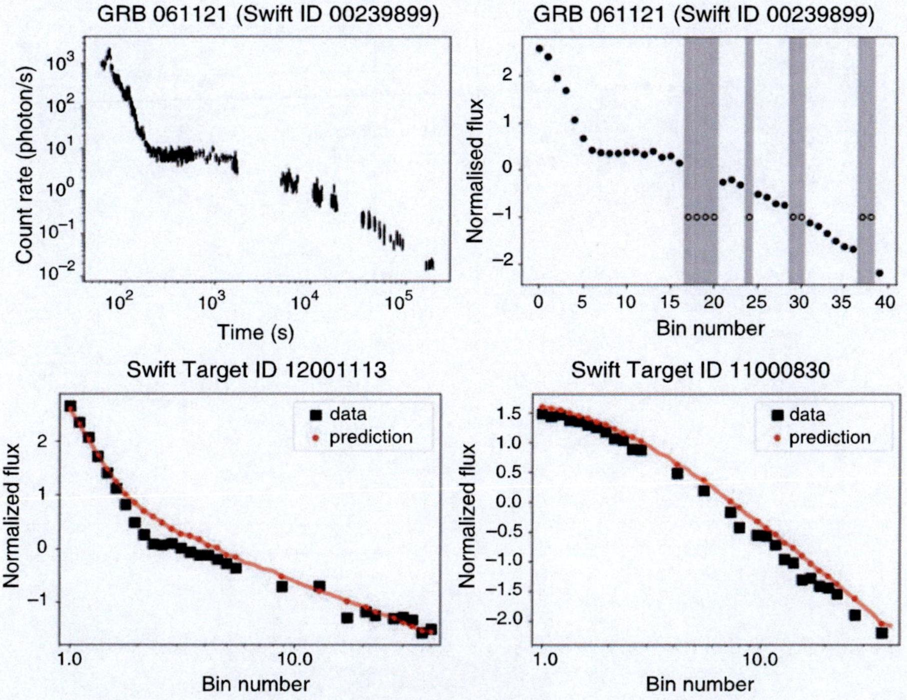

Fig. 1 *Top Panel:* results from our pre-processing procedure using the example of the canonical *Swift*-XRT light curve of GRB 061121 (*Swift* Trigger ID 00239899). Original *Swift* raw count rate light curve on the left. The pre-processing procedure preserves the light curve shape and aspect ratio (right). Values within empty bins (shaded regions) are set to -1 (unfilled data points); these points are excluded during training. *Bottom Panel:* Examples of light curve reconstructions

of the data, we employ the following model (ReLU activation unless specified otherwise):

- **Encoder:** Input layer: 40 inputs, dense layer 1: 40 neurons, dense layer 2: 24 neurons, bottleneck dense layer (linear activation): 2 latent dimensions
- **Decoder:** Input layer: 2 latent dimensions, dense layer 1: 24 neurons, dense layer 2: 40 neurons, output dense layer (linear activation): 40 outputs
- **Training:** train/test split: 80%/20%.
- **Stopping criterion:** validation loss with patience = 100 steps.

3 Results

We first validate our model against a simple 2-class (Ia: one break, steep first; Ib: one break, shallow first) synthetic light-curve sample. We find that our VAE model is able to generate realistic reconstructions (Fig. 1). Spectral clustering applied to the

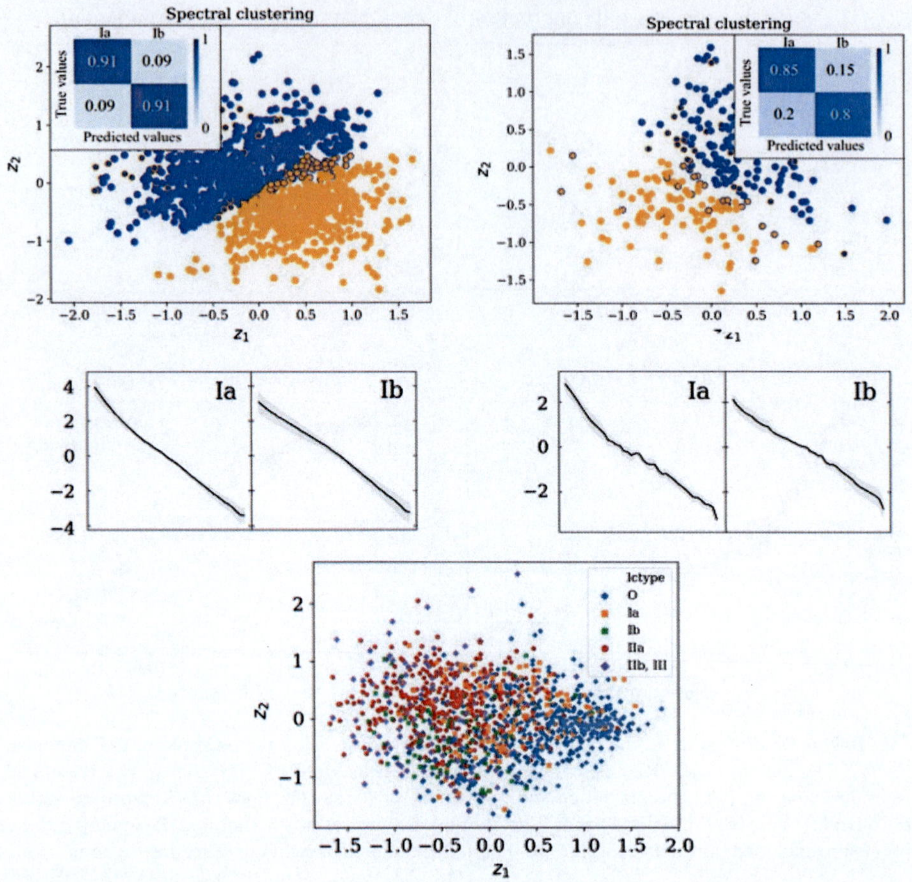

Fig. 2 *Above:* results for training performed on two-class synthetic data (left) and two-class real data (right). Top panels show the latent distribution of the data, with marker edge colors corresponding to true classes. Fill colors are a two-class prediction using spectral clustering. Bottom panels show average reconstructed light curves, with 1-sigma deviation around the mean (shaded region) for the predicted classes. *Left:* Latent distribution for the full real data set, colored by traditional classification

latent space recovers the input classes with very high accuracy (Fig. 2, left column). We generate the 'typical' light curve for each class by averaging the output of the decoder for each recovered class, which yields the expected shapes, thus validating the ability of our VAE model to map pre-processed *Swift* X-ray light curves to a 2D latent space. Applying our model to the sub-sample of real data that fall under Types Ia or Ib using the pre-computed labels from the classification available on the *Swift* website, we obtain similar results (Fig. 2, right column), indicating that our technique is suitable for real data, at least when restricted to light curves drawn from clearly segregated classes.

When carrying out embedding on the full real data set however, the clear separation is replaced by a rather more uniformly sampled latent space (Fig. 2, lower left). Categories that are separate in the traditional classification populate different regions of the latent space, but without separation, which signifies an absence of clustering in light curve space and suggests that light curves span a continuum of morphologies.

Improvements would involve further model testing, to ensure the absence of clear separation is not caused by our choice of model. VAEs are known to be sensitive to the balancing between reconstruction loss and regularisation, which could be ameliorated with a β-VAE. However, preliminary testing with more advanced models has not yielded significant improvements to latent space clustering, indicating that the lack of clear segregation may indeed be a real feature of the data.

References

1. Margutti, R., Zaninoni, E., Bernardini, et al.: The prompt-afterglow connection in gamma-ray bursts: a comprehensive statistical analysis of swift x-ray light curves. Month. Not. R. Astron. Soc. **428**(1), 729–742 (2013). https://doi.org/10.1093/mnras/sts066
2. Kingma, D.P., Welling, M.: Auto-encoding variational bayes (2013, e-prints). arXiv:1312.6114

Reconstructing Blended Galaxies with Machine Learning

Lavanya Nemani, Adriano Fontana, and Emiliano Merlin

Abstract Galaxy blending is a confusion effect created by the projection of photons from galaxies on the same line of sight, to the 2D plane. The upcoming deep extragalactic surveys like LSST and Euclid expect to see a blending fraction of up to 50% in the densest regions. For standard aperture photometry and for more complex techniques such as PSF-fitting and template-fitting algorithms, de-blending, the process of reconstructing the individual light profiles from blended sources, becomes crucial.

The current standard de-blending algorithms like SExtractor (Bertin and Arnouts, Astron Astrophys Suppl Ser 117(2):393–404, 1996) are based on threshold methods that simply assign each pixel to a single object, often failing to correctly take into account the real properties of the blended galaxies. With the advent of Machine Learning (ML) and Computer Vision in Astronomy we want to explore an unbiased and more accurate method of reconstructing individual light profiles using generative models.

1 Variational Auto-Encoders

Auto-Encoders [1] were first introduced as neural networks that are trained to reconstruct their inputs. They are generative models that consists of two parts: encoder and decoder. The encoder part of the model learns to reduce the high-dimensional input to an encoded representation and the decoder part learns how to reconstruct the input from the lower dimensional representation which is called the latent space or the bottleneck layer.

Variational Auto-Encoders [2], VAE, are a flavor of Auto-Encoders that use probablistic distribution for data generation, usually a normal distribution, such that instead of mapping the high-dimensional input to a fixed encoded representation, it

L. Nemani (✉) · A. Fontana · E. Merlin
INAF - Astronomical Observatory of Rome (OAR), Monte Porzio Catone, Italy
e-mail: lavanya.nemani@inaf.it; adriano.fontana@inaf.it; emiliano.merlin@inaf.it

F. Bufano et al. (eds.), *Machine Learning for Astrophysics*, Astrophysics and Space Science Proceedings 60, https://doi.org/10.1007/978-3-031-34167-0_18

is mapped to a distribution, giving us more control over how we want to model our latent distribution [3].

Two distinct VAE networks are needed to de-blend galaxies: One which learns how to reconstruct galaxy light profiles in isolation, and another one which uses the trained decoder (with fixed weights) of the first network to actually de-blend overlapping galaxies by reconstructing their individual light profiles [4]. The second network's encoder, that works on blended galaxy images, learns to map the high-dimensional input to a point in the latent space as if it was isolated over the training period. The latent space point can then be reconstructed to a full galaxy stamp using the first network's decoder which is already trained to reconstruct galaxy stamps in isolation.

2 Data Generation

To test the results of using this ML technique for de-blending we simulate stamps of galaxy images as expected from the Euclid survey in the VIS band. The galaxy images are built starting from a mock input catalogue created using EGG [5], a code that can generate mock galaxy catalogs with realistic positions, morphologies and fluxes. The catalog feeds the image simulation toolkit GalSim [6] to generate analytical double-Sersic profiles.

We also use Euclid*ized* realistic profiles that have been created as part of the Euclid Morphology Challenge [7, 8] using VAE's with Regressive Flows trained on HST images. The results on de-blending network trained on the realistic stamps are not discussed in this article and can be discussed with the author by email.

3 Results

The main focus of our work is to obtain accurate flux and morphology estimates for blended objects and clean light profiles to be used as priors for template fitting codes like T-PHOT [9]. We use several metrics to test our performance of the VAE algorithm comparing the reconstructed light profile of each simulated galaxy to the true input one, such as mean-square error and cosine distance. We also compare fluxes obtained on predicted images with the input catalogue fluxes. With our current best approach we are able to retrieve the original flux within 10% for 1σ as seen in Fig. 1 (whereas SExtractor [10] is within 16% for 1σ). In the future, we want to optimize the VAE's used for de-blending by implementing hyper-parameter tuning and additionally including data augmentation techniques for better performances.

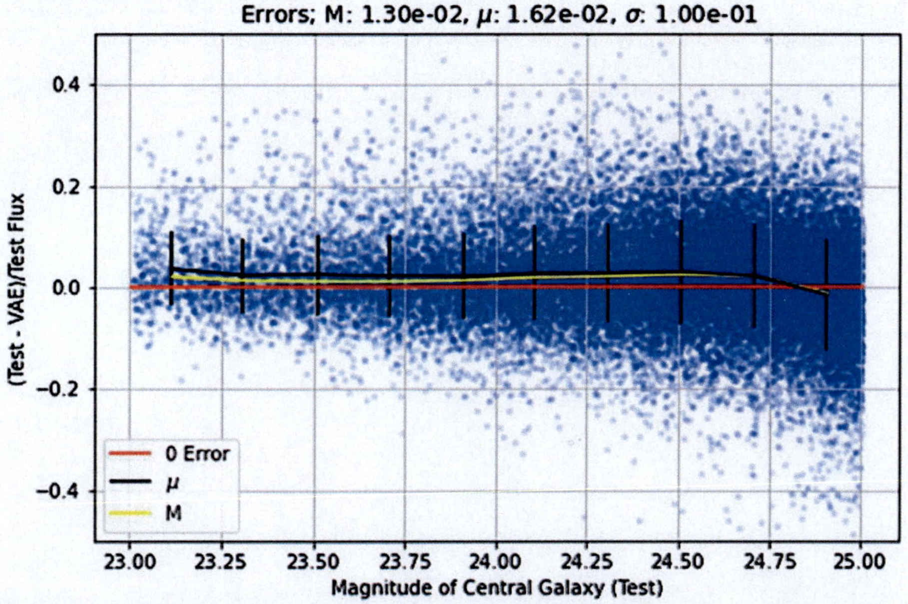

Fig. 1 Scaled error in flux of the prediction of the reconstruction of the central galaxy in the blended pair against its true magnitude, along with a binned mean, median and standard deviation. The flux is estimated by summing the pixels of the reconstructed galaxy stamp. Overall, the mean is 1.62e-2 and the std is 1.00e-1. The mean is very close to 0 for all galaxies with mag < 24.75 giving us largely unbiased results

References

1. Rumelhart, D.E., Hinton, G.E., Williams, R.J.: Learning internal representations by error propagation. Nature **323**(6088), 533–536 (1986).https://doi.org/10.1038/323533a0
2. Kingma, D.P., Welling, M.: Auto-encoding variational bayes (2015, preprint). https://doi.org/10.48550/arXiv.1312.6114
3. Higgins, I., Matthey, L., et al.: Early visual concept learning with unsupervised deep learning (2016, preprint). https://doi.org/10.48550/arXiv.1606.05579
4. Arcelin, B., Doux, C., et al.: Deblending galaxies with variational autoencoders: a joint multiband, multi-instrument approach. Month. Not. R. Astron. Soc. **500**(1), 531–547 (2020). https://doi.org/10.1093/mnras/staa3062
5. Schreiber, C., et al.: EGG: hatching a mock Universe from empirical prescriptions. Astron. Astrophys. **602**, A96 (2017). https://doi.org/10.1051/0004-6361/201629123
6. Rowe, B., Jarvis, M., et al.: GalSim: the modular galaxy image simulation toolkit. Astron. Comput. **10**, 121–150 (2014). https://doi.org/10.1016/j.ascom.2015.02.002
7. Bretonnière, H., et al.: Euclid preparation: XIII. Forecasts for galaxy morphology with the euclid survey using deep generative models. Astron. Astrophys. **657**, A90 (2022). https://doi.org/10.1051/0004-6361/202141393
8. Merlin, E., et al.: Euclid preparation. XXV. The euclid morphology challenge–towards model-fitting photometry for billions of galaxies (2022, preprint)

9. Merlin, E., et al.: T-PHOT: a new code for PSF-matched, prior-based, multiwavelength extragalactic deconfusion photometry. Astron. Astrophys. **582**, A15 (2015). https://doi.org/10.1051/0004-6361/201526471
10. Bertin, E., Arnouts, S.: SExtractor: software for source extraction. Astron. Astrophys. Suppl. Ser. **117**(2), 393–404 (1996). https://doi.org/10.1051/aas:1996164

Time Domain Astroinformatics

Massimo Brescia

Abstract Astronomy has entered the multi-messenger data era and Data Science has found widespread use in a large variety of applications. The exploitation of synoptic (multi-band and multi-epoch) surveys, like Rubin-LSST, requires an extensive use of automatic methods for data processing and interpretation. With data volumes in the petabyte domain, the discrimination of time critical information has already exceeded the capabilities of human operators and crowds of scientists have extreme difficulty to manage such amounts of data in multi-dimensional domains. Therefore, a change of perspective is required to explore Time Domain Astronomy (TDA) in an efficient and semi-automatic way, introducing the Time Domain Astroinformatics as the virtuous synergy between TDA and Data Science domains.

1 Introduction

Traditional Astronomy is faced on the 3D hyper-surface of the past light cone in the 4D spacetime. In such a context, Time Domain Astronomy (TDA) describes a 4D hypervolume as we move along the time axis of the 4D spacetime. The baseline concept of a light curve shows the profile brightness of a sky source that changes with time. As known, there are two main reasons why this happens: *(i) intrinsic variation*, that is happening inside the source itself, like in the case of Cepheids, Supernovae and RR Lyrae, by changing some of its physical features, like dimensions and temperature; *(ii) extrinsic variation*, that is happening outside the source, by changing the amount of visible light, for example in the case of exoplanet transits and eclipsing binaries.

As known, the phase light curve diagram is a very useful way to understand the shape of the light curve and to verify the periodicity guessed. And there are several

M. Brescia (✉)
Department of Physics "E. Pancini", University of Napoli Federico II, Napoli, Italy
e-mail: massimo.brescia@unina.it

F. Bufano et al. (eds.), *Machine Learning for Astrophysics*, Astrophysics and Space Science Proceedings 60, https://doi.org/10.1007/978-3-031-34167-0_19

methods to make a realistic guess about the period, using the magnitude information or through the Fourier analysis in the frequency domain [1].

Another well-known method to calculate the true period of a light curve is to search for periodic signals in the light curve. This can be done by computing the Fourier spectrum from the light curve. A classic Fourier transform analysis requires the data points to be evenly spaced in time, which is not likely to happen for astronomical observations. A type of Fourier transform was then developed for unevenly-spaced data, called Lomb-Scargle periodogram. Functionally, the periodogram can be used in the same way, to indicate powerful frequencies in the light curve, and it is a very effective and quick way to recognize the source from its derived phased light curve [2]. However, the analysis of a periodogram in the case of irregular sampling is often limited by the possibility for fully fixing its statistical properties. Often the experimental conditions do not permit a regular sampling of signals and to define a set of natural frequencies for which to compute the periodogram. Furthermore, in the case of irregular sampling the periodogram is not equivalent to the least-squares fit of sinusoidal functions. Therefore, the periodogram becomes a simple statistical measure of similarity between the experimental time series and a discrete sinusoidal signal of frequency. An additional difficulty introduced by an irregular sampling is the lack of computing efficiency of the algorithm. Some more complex solutions have been proposed, such as in [3], where a rebinning of the signal makes possible to transform the light curve in a regular time series.

So far, we understood also an important aspect: the light curve, at the end, is a time series and to that we could apply the theory of signals. There are various properties or components that influence decisions in the data science pipeline for the development of forecasting/classification systems. Some of them are very useful in the case of a light curve analysis. For example, after removing the more shape and periodic components (trend, seasonality, cyclic patterns) from the time series, the remaining part is known as the irregular component or residuals. Residuals are not explainable by any trend, seasonal or cyclic behaviour, which can still cause an impact on the dynamics of the time series. At any given point in time, a time series can be decomposed in an additive way into the above-mentioned components. Moreover, the autocorrelation of a time series quantifies this dependency according to the correlation of each observation with its past values. This property provides important structural information about the series. For example, in case of a low autocorrelation of values along the temporal sequence, the series is referred to as white noise.

Most of the data stream properties assume that the underlying time series is represented by a single dimension, thus referred to as univariate. Whenever a time series has a multi-dimensional shape is referred to as a multivariate time series. The additional variables in multivariate time series can be used as explanatory information when modelling a particular target variable of the time series [4].

Time series forecasting is a difficult problem with no easy answer. Although there are countless statistical models that claim to outperform each other, it is never clear yet which model is best. That being said, ARMA-based models (Auto-Regressive

Moving Average) can achieve some scores on most time-series problems and are well-suited as a baseline model in any time series problem [5].

2 Synoptic Sky Surveys

The problem of detection and classification of variable sources becomes more complex when we want to dynamically explore wide portions of sky. This is the case of nowadays trend to develop the so-called Synoptic digital sky surveys, a panoramic cosmic cinematography, which are now the dominant data producers in astronomy. Their scale goes from Terascale to Petascale data streams. Of course this is a major new growth area of astrophysics driven by the new generation of large digital synoptic sky surveys (CRTS, PTF/ZTF, PanSTARRS, SkyMapper, to quote just a few), leading to Rubin-LSST and SKA. They represent a broader significance for an automated, real-time knowledge discovery in massive data streams, but at the same time they pose astronomers in front of more challenging efforts to efficiently explore millions of sources in real time. For example, when the Vera C. Rubin Observatory Legacy Survey of Space and Time (Rubin-LSST) will be operative, just about the 0.1% of all detected transients will be spectroscopically classified. This pose a very challenging problem to make a massive and optimized use of classification methods in the photometric domain [6].

Furthermore, traditionally, Astronomy was a data-starved science. Typical research and analysis methods were shaped by this environment. Now surveys are changing such a vision; data is becoming abundant and of unprecedented quality. And in particular, it is characterized by a high-dimensionality parameter space, well beyond the limited 3D space normally understandable by humans. For example, Rubin-LSST will deliver positions, magnitudes and variability information for virtually everything in the southern sky to 24th–27th magnitude, with better controlled systematics than current surveys.

As known a synoptic sky survey can be designed by taking into account several instrument and environment features. For sure, one of the most important is the depth of the survey. In practice it means to correlate the effective collecting area of the telescope, the duration of the exposures and the overall efficiency of the whole instrument, normalized by the astronomical quality of the site or the atmospheric seeing. The survey should be also calibrated in terms of its potential capability to perform the scientific discoveries, which means to take into account the amount of exposures for each pointing, the covered area, the spectral range and its depth.

We must always consider that the ultimate deliverable of an astronomical facility is not the telescope or the focal plane instruments, but it is the fully reduced data. But to produce them, we must extract all features from images and catalogues, or in other words, we must define the parameter space (PS).

In the case of TDA, the multi-dimensional PS is composed by not only the usual features, like flux, wavelength, coordinates, but also specific parameters for light curves, such as the significance of periodicity, the power law, the amplitudes and their statistical descriptors. And for surveys, the total exposures as well as the characterization of the specific cadence strategy. TDA surveys generate thousands time-domain events per observation, including everything that changed with respect to deep templates (variables, explosive transients, asteroids, etc.). A first example was the Palomar Quest Event Factory, (the precursor for the Palomar Transient Facility and the Catalina Real-Time Survey), the first to perform real-time detection and publishing of transients using the Virtual Observatory paradigm to generate follow-up alerts for transients. The Zwicky Transient Facility [7], is a survey aimed at obtaining a systematic study of the optical night sky, being able to scan the entire Northern sky every 2 days. Up to the incoming Rubin-LSST, a wide (half the sky), deep (24.5/27.5 mag) and fast (image the sky once every 3 days) survey, which since the 2024 will repeatedly image the sky for 10 years. It will produce a stream of about 10 million time-domain events per night, detected and transmitted within 60 s from observation. The final products will be a catalogue of orbits for ~6 million bodies in the Solar System, about 20 billion galaxies, 17 billion stars and about 7 trillion single-epoch detections [8].

In order to be able to process such events, it is required to rapidly analyse data streams and to generate the so-called transient alerts for an immediate follow-up with other observing facilities. Such a system should be able to plan, measure and transmit information about position, PSF flux, aperture fluxes, trailed model fits, shape, light curves in all bands, variability characterization as well as cutouts centered on the object and templates for the difference imaging [9]. If we multiply such an information package for the billions of detected sources along the lifetime of a survey, we easily can insert the TDA in a Big Data scenario.

3 TDA as a Big Data Challenge

From a conceptual viewpoint, a Big Data system is usually defined as a data ecosystem characterized by the so-called five-V rule: Volume, Variety, Velocity, Veracity and Value. TDA, with the plethora of available and incoming wide surveys, is now entered in such kind of data domain. This means two things: first, from a statistical point of view, as based on big data, its parameter space contains an almost complete set of information required to provide answers to all its scientific questions. But they are hidden within the massive amount of data. Second, it requires a change of perspective for its efficient analysis and exploration, which should be based on the paradigms of self-adaptive and automatic methods belonging to the area of machine learning paradigms.

Classification of transient events is essential for their astrophysical interpretation and uses. But in a data-intensive environment, like modern surveys of variable sky, the problem is extremely challenging. The pre-classification must be done in real time and iterated dynamically, the human classification is already unsustainable, and will not scale to the Petascale data streams, data are sparse and heterogeneous, the best trade-off between completeness and contamination must be found, follow-up resources are expensive and/or limited and it should be dedicated only to the most interesting events, the offline classification should be used to dynamically refine the classification as new data come in. Moreover, traditional pipelines do not capture a lot of the relevant contextual information, such as the prior/expert knowledge. The event classification in all current or incoming surveys are (and must be) based on a virtuous loop, based on essential incorporation of the contextual information (archival and from data themselves), automated selection of follow-up observations, given the available resources and their cost, massive exploitation of machine learning paradigms to obtain an efficient dynamical and iterative system for event detection and characterization.

A basic principle is that from a light curve it is possible to derive dozens of parameters and statistical measures. This turns heterogeneous light curves into homogeneous feature vectors in the parameter space. Such vectors can be applied to a variety of automated classification methods, derived from the paradigm of Machine and Deep Learning.

For example, in a past experiment [10], using the knowledge base extracted from the Catalina sky survey, the scope of the tests was the analysis of the parameter space, in order to identify the relevant features defining any light curve object, followed by the classification of different variable types through Machine Learning methodology. The promising results however, confirm the need of a larger amount of data with known class to optimize the supervised classification process. In particular, such kind of experiments provided the requirement to surveys like LSST, enabling the possibility to have available a data sample large and unbiased, to equally sample all evolutionary aspects and in some cases, such as for Supernova subtypes, to relate statistically the sources with the features (age and metallicity) of their progenitors.

Besides a vast literature related to the application of Machine Learning methods for photometric classification of variable sources [11–13], it is important to mention a deep learning model, particularly useful in time series prediction problems, like reconstruction of light curves for SNe and other kinds of transients. The model is called LSTM or Long-Short-Term-Memory [14], and is a hierarchical series of layers, like synthetic memory cells, each one including special functional gates, able to correlate and maintain memory of the past sequence of the signal, providing as output the prediction of the signal trend in the future. It is a special case of so-called recurrent learning models, able to overcome the traditional problem of the lack of long-term memory for machine learning hierarchical models. One of the first published works demonstrated the high performance capability for LSTM to recognize light curve profiles and to classify the variable sources.

Among several proposed variations of the LSTM model, there was the so-called Convolutional LSTM, which replaces the fully-connected layers in the canonical LSTM with convolutional layers, one of the most powerful was the Temporal Convolutional Network (TCM), which leveraged convolutional networks to model sequence-based data. Naturally, it was demonstrated to be an ideal candidate for time series analysis tasks [15]. Its key concept is the dilated convolution. They extend the concept of introducing padding elements to input images, a common technique used in CNNs to solve the problem of involving lateral sides of an image within the dynamical convolution process, by padding also the input sequences of a time series. This makes able the model to detect dependencies between items that are not only close to each other, but also at entirely different positions.

A TCN was for example proposed as a generative model of SNIa given a partial light curve [16]. In this case the DNN was specifically designed to predict one time-step in the future and can acts as a good generative system of synthetic light curves to be used as learning knowledge base. To obtain a sequence of predictions, they feed in the predicted values back into the DNN as if they were part of the observations.

Kruk et al. [17] proposed a method based on reinforcement learning, by rewarding desired behaviours and punishing negative behaviours. This method assigns positive or negative values to the agent's actions, depending on the desired or undesired behaviour achieved. This process allows to seek the maximum long-term overall reward towards the optimal solution. These long-term goals help prevent the agent from stalling on lesser goals. With time, the agent learns to avoid the negative and seek the positive. This learning method has been adopted in artificial intelligence (AI) as a way of directing unsupervised machine learning through rewards and penalties.

4 Conclusions

Virtually all scientific disciplines, in particular Astrophysics, have undergone a drastic change of perspective in the way they manage the cognitive investigation, having to face new challenges, such as the management of large and complex datasets and their exploration to deduce new knowledge.

This is why the modern scientific discovery process must follow a virtuous processing cycle, focused on collecting, storing, retrieving and mining data, trying to find hidden correlations, detect anomalies and analyse outliers, in order to understand the underlying laws that may provide new knowledge. Especially in TDA projects such as SKA or Rubin-LSST this approach is fundamental, or rather, data triage becomes the only means capable of not being overwhelmed by the amount of data available. The increased complexity and the challenges to be faced implied a new methodological approach: from hypothesis-based science to data-driven science. Both approaches can be considered complementary, but the latter increases the crucial role of the steps that directly involve data management and analysis, starting from the fundamental assumption that all answers to scientific questions are nested within the data and therefore the data itself must guide scientific research, obviously in the most automatic possible way.

The virtuous data exploration cycle in TDA should be hence based on a double interpretation of the concept of Distributed Data Mining [18]: first, *distributed mining of data*, which means to move code to data, where intelligent methods can operate on the whole data system. Second, *mining of distributed data*, where specific data portions are distributed worldwide through accredited data brokers to enable specific science use cases. Such dichotomy is the most effective solution to really extract the best knowledge with Time Domain Astroinformatics.

References

1. Ripley, B.D.: Statistical Inference for Spatial Processes. University Press, Cambridge (1988). https://doi.org/10.1017/CBO9780511624131
2. Barret & Vaughan: Maximum likelihood fitting of X-ray power density spectra: application to high-frequency quasi-periodic oscillations from the neutron star X-ray binary 4U1608-522. ApJ. **746**(2), 131 (2012). https://doi.org/10.1088/0004-637X/746/2/131
3. Vio, R., et al.: Irregular time series in astronomy and the use of the Lomb-Scargle periodogram. Astron. Comput. **1**, 5–16 (2013). https://doi.org/10.1016/j.ascom.2012.12.001
4. Ivezić, Ž., VanderPlas, J.T.: Periodograms for multiband astronomical time series. Commun. Konkoly Observatory. **105**, 97–100 (2016)
5. Box, G.E.P., et al.: Time Series Analysis: Forecasting and Control, 5th edn, p. 53. ISBN 978-1-118-67492-5. Wiley, Hoboken, NJ (2016)
6. Ivezić, Ž., et al.: LSST: from science drivers to reference design and anticipated data products. ApJ. **873**(2), 111 (2019). https://doi.org/10.3847/1538-4357/ab042c
7. Graham, M.L., et al.: The Zwicky transient facility: science objectives. PASP. **131**, 1001 (2019). https://doi.org/10.1088/1538-3873/ab006c
8. Li, X., et al.: Preparing to discover the unknown with Rubin LSST: time domain. ApJS. **258**, 1 (2021). https://doi.org/10.3847/1538-4365/ac3bca
9. Goldstein, D.A., et al.: Automated transient identification in the dark energy survey. AJ. **150**, 82 (2015). https://doi.org/10.1088/0004-6256/150/3/82
10. D'Isanto, A., et al.: An analysis of feature relevance in the classification of astronomical transients with machine learning methods. MNRAS. **457**, 3 (2016). https://doi.org/10.1093/mnras/stw157
11. Charnock, T., Moss, A.: Deep recurrent neural networks for supernovae classification. ApJ. **837**(2), L28 (2017). https://doi.org/10.3847/2041-8213/aa603d
12. Vicedomini, M., et al.: Statistical characterization and classification of astronomical transients with machine learning in the era of the Vera C. Rubin Observatory. Intelligent astrophysics. In: Zelinka, I., Brescia, M., Baron, D. (eds.) Emergence, Complexity and Computation, vol. 39, pp. 81–113. Springer, Cham (2021) ISBN: 978-3-030-65867-0
13. Hsu, B., et al.: Photometrically-classified superluminous supernovae from the pan-STARRS1 medium deep survey: a case study for science with machine learning-based classification. Submitted to ApJ. eprint arXiv:2204.09809 (2022)
14. Hochreiter, S., Schmidhuber, J.: Long short-term memory. Neural Comput. **9**, 8 (1997). https://doi.org/10.1162/neco.1997.9.8.1735
15. Bai, S., et al.: An empirical evaluation of generic convolutional and recurrent networks for sequence modeling. e-print arXiv:1803.01271. (2018). https://doi.org/10.48550/arXiv.1803.01271
16. Muthukrishna, D.: Real-time photometric identification of transients. American Astronomical Society meeting #237, id. 509.04. Bull. Am. Astron. Soc. **53**(1), e-id 2021n1i509p04 (2020)
17. Kruk, S., et al.: Hubble Asteroid Hunter. I. Identifying asteroid trails in Hubble Space Telescope images. A&A. **661**, A85 (2022). https://doi.org/10.1051/0004-6361/202142998
18. Giannella, C., et al.: Distributed data mining for astronomy catalogs. Proceedings of 9th Workshop on Mining Scientific and Engineering Datasets, as part of the SIAM International Conference on Data Mining (SDM) (2006)

A Convolutional Neural Network to Characterise the Internal Structure of Stars

J. C. Suárez, A. García Hernández, R. Maestre, S. Barceló Forteza, and G. Mirouh

Abstract In this work we used a convolutional neural network to determine the so called, large separation ($\Delta\nu$), a pattern in the oscillation frequency that is a proxy for the stellar mean density, which is key to constraint the internal structure of stars, and thereby their evolution. We trained a CNN with a grid of asteroseismic models representative of intermediate-mass stars (from 1.5 to 3 times more massive than the Sun) and validated it with a sample of observed benchmark stars. We found that the CNN was around 56% accurate finding the actual values of $\Delta\nu$, and 94% when considering multiples and/or submultiples of $\Delta\nu$. This method will allow us to analyse massively thousands of A-F stars observed by past, present and future space missions such as CoRoT, Kepler/K2, TESS, and the upcoming PLATO mission.

1 The Context

Asteroseismology is nowadays the most powerful tool to probe the interiors of stars, allowing us to test theories on stellar structure and evolution. However, this technique relies on the identification of hundreds of oscillation modes, a very challenging and uncertain task mainly because of model degeneracy. Some theoretical works predicted the existence of a quasiperiodicity that scales with the stellar mean density for intermediate-mass stars, such as the well-known large separation [8, 9], then confirmed empirically by García Hernández et al. [4]. This opened the door to multi-variable correlation analyses [3, 6], and the study of the rotational splitting [7]. Despite all this progress, the robust and unambiguous identification of the patterns still requires the experimented eye of the researcher.

J. C. Suárez (✉) · A. García Hernández · R. Maestre · S. Barceló Forteza · G. Mirouh
Department of Física Teórica y del Cosmos, Universidad de Granada, Granada, Spain
e-mail: jcsuarez@ugr.es

F. Bufano et al. (eds.), *Machine Learning for Astrophysics*, Astrophysics and Space Science Proceedings 60, https://doi.org/10.1007/978-3-031-34167-0_20

2 The CNN

In order to minimise subjective criteria in such a pattern recognition, we trained a convolutional neural network (CNN from now on) to identify the so-called large separation, $\Delta\nu$, in δ Scuti stars.[1] Each NN consists in the following stacked layers: convolution, max pooling, and batch normalisation with dropout. These are then combined into a dense layer with a *Softmax* activation function that normalises the output into a probability distribution i.e. representing a categorical distribution. Each convolution layer of the CNN is fed with a three-dimensional vector composed by the autocorrelation function (AC), the histogram of frequency differences (HDF) and the discrete Fourier transform (DFT), all of them providing an estimate of $\Delta\nu$ from asteroseismic models. The dataset was built with 500k asteroseismic models (equilibrium and oscillation frequencies), where mass, metallicity and initial rotation velocity were varied as to have a dense grid of representative models of intermediate-mass stars. Following [9], for each model we considered a frequency range covering radial orders from $n = 2$ to $n = 8$, from which the 30 highest-amplitude modes were kept. The amplitude was simulated assuming a visibility law uniquely based on the photometric cancellation effect (up to $\ell = 3$), where the mode visibility decrease for increasing ℓ in the form $A_\ell \sim \ell^{-1/2}$.

We trained a CNN to provide a probability over the C classes, where C is a partition of mesh p in the interval [0, 100] μHz, in which $\Delta\nu$ is typically observed for the studied stars. We tested the *internal* accuracy of the CNN against a control set composed by around 50k models randomly (uniform) distributed along the model grid to avoid any selection bias (Fig. 1, left). We then evaluated the top-k categorical accuracy rate, meaning the success when the target class is within the k predictions

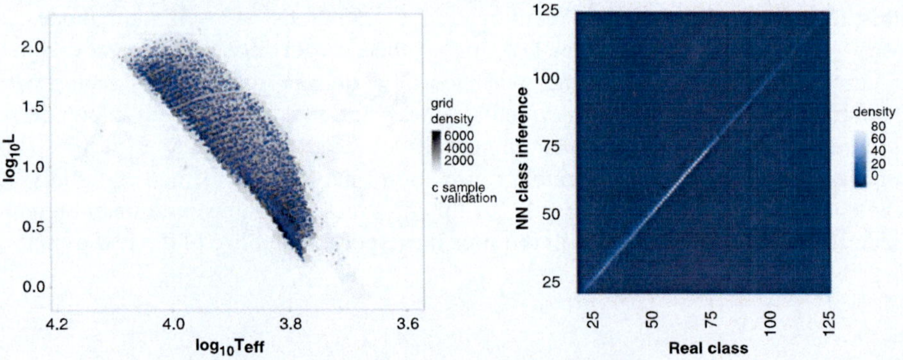

Fig. 1 Left: Hertzprung-Russel diagram of the model grid used in this work represented as 2D density. Blue dots represent the 50k models used as control sample. Right panel: $\Delta\nu$ confusion matrix for the control sample (97% of accuracy for top-1 ranked probabilities)

[1] A type of intermediate-mass that pulsate with pressure modes.

ranked by probability. For the control sample, an accuracy of 97% is reached for top-1 probabilities (Fig. 1, right).

3 Results and Preliminary Conclusions

We applied the CNN to a benchmark sample composed of 11 eclipsing binary systems with a δ Scuti star component (see details in [5]), for which independent measures of $\Delta\nu$ and mean density are available. The confusion matrix (Fig. 2) shows that for about 56% of the stars the CNN precisely found the correct value of the large separation (1:1 line), using top-1 inferences. This gives a RMSE of 28.09 muHz for the complete set of stars. Interestingly, the rest of points are not randomly distributed in the diagram, but around multiples and submultiples of the 1:1 relation. When these multiples are considered the RMSE of the match significantly decreases to 4.80 muHz. This implies that the CNN is properly finding $\Delta\nu$ in all the cases, but is not able to *decide* whether is the actual $\Delta\nu$ value or a multiple/submultiple. This problem comes from the lack of information about the visibility of the modes, which, in the range of the observed frequencies, may be prone to display submultiple/multiple instead the actual $\Delta\nu$.

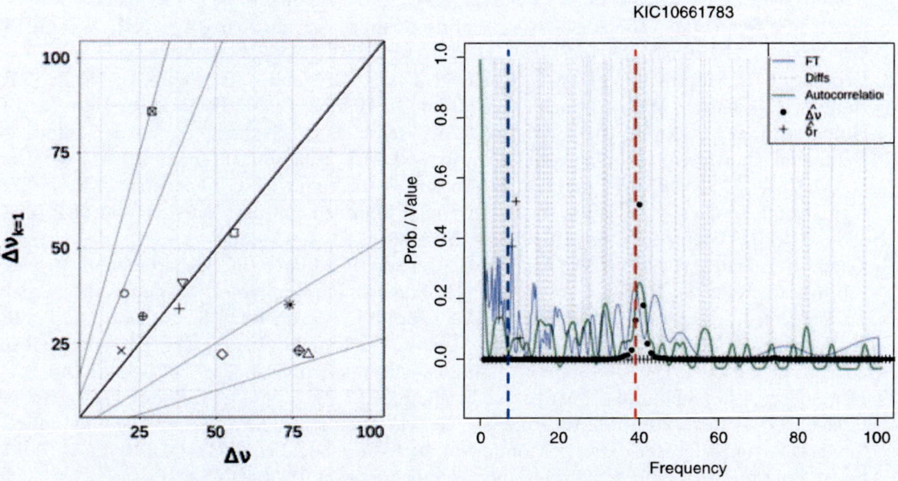

Fig. 2 Left panel: comparison of $\Delta\nu$ inferred by the CNN for the benchmark sample of δ Scuti stars in eclipsing binary systems, with those obtained with other methods [5]. Each symbol represent a different star of the sample. Lines show the 1:1, 0.5:2, and 0.33:3 match relations. Right panel: CNN inference diagram for the star KIC 10661783. Black dots represent the CNN inference given by a probability density as a function of the frequency bucket (large separation). For illustration the DFT, AC and HFD are also depicted. Vertical red line represents the actual value (measured independently). For illustration, the location of the rotational splitting frequency is also depicted as a blue dashed line

The CNN was trained with appropriate asteroseismic models to eliminate the human factor for estimating $\Delta\nu$, a proxy of the mean density and surface gravity of intermediate-mass stars. Nonetheless, this preliminary conclusion must be considered with caution until (1) additional tests are performed with a large sample of stars, and (2) additional information is considered to brake degeneracy of solutions with multiples of the $\Delta\nu$, e.g. corrections for gravity darkening effects [1], precise luminosities from Gaia mission, or constraints from other seismic indices [2].

Acknowledgments JCS, AGH, RM, SBF, and GM acknowledge support by University of Granada and by Spanish public funds for research under project "Contribution of the UGR to the PLATO2.0 space mission. Phases C/D-1", funded by MCNI/AEI/PID2019-107061GB-C64. AGH also acknowledges support from "FEDER/Junta de Andalucía-Consejería de Economía y Conocimiento" under project E-FQM-041-UGR18 by Universidad de Granada.

References

1. Barceló Forteza, S., Moya, A., Barrado, D., Solano, E., Martín-Ruiz, S., Suárez, J.C., García Hernández, A.: Unveiling the power spectra of δ Scuti stars with TESS. The temperature, gravity, and frequency scaling relation. Astron. Astrophys. **638**, A59 (2020). https://doi.org/10.1051/0004-6361/201937262
2. Barceló Forteza, S., Roca Cortés, T., García, R.A.: The envelope of the power spectra of over a thousand δ Scuti stars. The $\bar{T}_{eff}$ - ν_{max} scaling relation. Astron. Astrophys. **614**, A46 (2018). https://doi.org/10.1051/0004-6361/201731803
3. Bedding, T.R., Murphy, S.J., Hey, D.R., Huber, D., Li, T., Smalley, B., Stello, D., White, T.R., Ball, W.H., Chaplin, W.J., Colman, I.L., Fuller, J., Gaidos, E., Harbeck, D.R., Hermes, J.J., Holdsworth, D.L., Li, G., Li, Y., Mann, A.W., Reese, D.R., Sekaran, S., Yu, J., Antoci, V., Bergmann, C., Brown, T.M., Howard, A.W., Ireland, M.J., Isaacson, H., Jenkins, J.M., Kjeldsen, H., McCully, C., Rabus, M., Rains, A.D., Ricker, G.R., Tinney, C.G., Vanderspek, R.K.: Very regular high-frequency pulsation modes in young intermediate-mass stars. Nature **581**(7807), 147–151 (2020). https://doi.org/10.1038/s41586-020-2226-8
4. García Hernández, A., Moya, A., Michel, E., Garrido, R., Suárez, J.C., Rodríguez, E., Amado, P.J., Martín-Ruiz, S., Rolland, A., Poretti, E., Samadi, R., Baglin, A., Auvergne, M., Catala, C., Lefevre, L., Baudin, F., Rodriguez, E., Amado, P.J., Martin-Ruiz, S., Rolland, A., Poretti, E., Samadi, R., Baglin, A., Auvergne, M., Catala, C., Lefevre, L., Baudin, F.: Asteroseismic analysis of the CoRoT δ Scuti star HD 174936. Astron. Astrophys. **506**(1), 79–83 (2009)
5. Hernández, A.G., Suárez, J.C., Moya, A., Monteiro, M.J.P.F.G., Guo, Z., Reese, D.R., Pascual-Granado, J., Forteza, S.B., Martín-Ruiz, S., Garrido, R., Nieto, J.: Precise surface gravities of δ Scuti stars from asteroseismology. Month. Not. R. Astron. Soc. Lett. **471**(1), L140–L144 (2017). https://doi.org/10.1093/mnrasl/slx117. http://arxiv.org/abs/1707.06835
6. Moya, A., Suárez, J.C., García Hernández, A., Mendoza, M.A.: Semi-empirical seismic relations of A-F stars from CoRoT and Kepler legacy data. Month. Not. R. Astron. Soc. **471**(2), 2491–2497 (2017). https://doi.org/10.1093/mnras/stx1717. https://arxiv.org/pdf/1707.02082.pdf
7. Ramón-Ballesta, A., García Hernández, A., Suárez, J.C., Rodón, J.R., Pascual-Granado, J., Garrido, R.: Study of rotational splittings in δ Scuti stars using pattern finding techniques. Month. Not. R. Astron. Soc. **505**(4), 6217–6224 (2021). https://doi.org/10.1093/mnras/stab1719

8. Reese, D., Lignières, F., Rieutord, M.: Regular patterns in the acoustic spectrum of rapidly rotating stars. Astron. Astrophys. **481**(2), 449–452 (2008)
9. Suárez, J.C., Hernández, A.G., Moya, A., Rodrigo, C., Solano, E., Garrido, R., Rodón, J.R.: Measuring mean densities of delta Scuti stars with asteroseismology. Theoretical properties of large separations using TOUCAN. Astron. Astrophys. **563**(A7), 11 (2014)

Finding Stellar Flares with Recurrent Deep Neural Networks

Attila Bódi and Krisztián Vida

Abstract At present, low-mass, cool M dwarfs are the prime targets of planet searches, since the habitable zone is much closer to the central object in cool stars than in the case of a solar-like star; thus, detecting a possibly habitable Earth-like planet is easier. However, the late spectral type of these stars and the magnetic activity associated with these could pose a threat to habitability. To study this threat, we have to find stellar flares that are an important tracers of magnetic activity, but automatically and accurately finding them is still a challenge to researchers in the Big Data era of astronomy. Here we present an experiment to detect flares in space-borne photometric data using deep neural networks. Using a set of artificial data and real photometric data we trained a set of neural networks, and found that the best performing architectures were the recurrent neural networks (RNNs) using Long Short-Term Memory (LSTM) layers. The aim for the trained network is not just detect flares but also be able to distinguish typical false signals (e.g. maxima of RR Lyr stars) from real flares. The best trained network detected flares over 5σ with $\geq$80% recall and precision. Testing the network—trained on Kepler space telescope observations—on Transiting Exoplanet Survey Satellite (TESS) light curves showed that the neural net is able to generalize and find flares with similar effectiveness in completely new data with different sampling and characteristics from those of the training set.

A. Bódi (✉) · K. Vida
Konkoly Observatory, Research Centre for Astronomy and Earth Sciences, Eötvös Loránd Research Network, Budapest, Hungary

CSFK, MTA Centre of Excellence, Budapest, Hungary
e-mail: bodi.attila@csfk.org; vida.krisztian@csfk.org

F. Bufano et al. (eds.), *Machine Learning for Astrophysics*, Astrophysics and Space Science Proceedings 60, https://doi.org/10.1007/978-3-031-34167-0_21

1 Introduction

In astronomy and astrobiology, the most common definition of an exoplanet's habitable zone (HZ) is based on the incoming flux from its host star. HZ is a range of orbits around a star within which liquid water can be found on the surface of a planet given sufficient atmospheric pressure. However, several other factors may play a significant role in planetary habitability, such as stellar activity, planetary atmosphere and magnetism, and planetary surface conditions.

To maximize the chances of finding potentially habitable Earth-like planets, the best targets to be observed are M-dwarf stars for different reasons: first they are the most common stars in our Galaxy, then the planets orbiting them have the shortest orbital period and finally the stronger relative signal makes the transit detection easier. Recent studies suggested that more than 80% of M-dwarf stars should have at least one Earth-sized planet [1].

On the other hand, in searching for life, we should take into account that M dwarfs are active [2, 3], which may decrease the possibility of habitability even if the planet is in the HZ. A large number of energetic transients can erode/destroy planetary atmospheres over time [4, 5]. But activity can also support life as UV radiation is needed for abiogenesis, which could be provided by flares around M-dwarfs [6]. Thus, there must be a sweet spot regarding the amount of flares, where the radiation does not yet destroy the planetary atmosphere, and can initiate abiogenesis [7].

The role of stellar activity in planetary habitability has motivated our research to exploit the power of recurrent deep learning methods to boost the search for stellar flares and to improve our understanding of this field.

2 Flare Detection with Recurrent Deep Neural Networks

Every deep learning method requires a training sample. We used space-based, quasi-continuous, one-minute-cadence light curves observed by the Kepler space telescopes to build a reliable data set. Before training, shorter gaps were filled by random noise characteristic of the quiescent part of the given data set and each light curve has been re-sampled and normalized. Flaring events were labeled by the `FLATW'RM` code,[1] then vetted and–if needed–re-flagged manually. The sample was enriched with stars that show flare-like variations, which are considered astrophysical noise. A typical flare event lasts about tens of minutes to an hour and shows steep brightening followed by a shallower fading/decay phase (see Fig. 1).

We designed and tested several neural network architectures and using different metrics, an RNN consisting of three LSTM and two dropout layers has been selected

[1] https://github.com/vidakris/flatwrm[8].

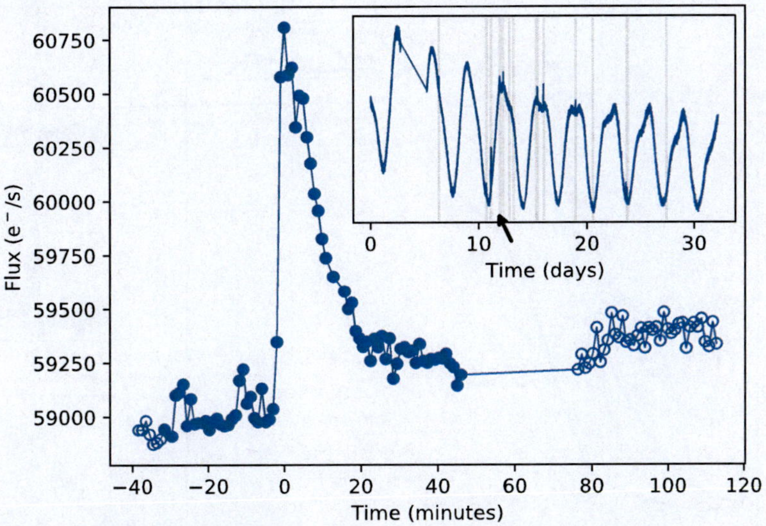

Fig. 1 Short cadence Kepler light curve of KIC 1570924. The inset shows the full data set, where the flares are labeled by vertical light gray bars. A significant event is highlighted. The filled points are flagged by the RNN as part of the event. Note that some pre-main-event flares can also be seen, which were detected as well

as the best performing one [9]. The precision and recall values as the function of flare signal-to-noise ratio can be seen in the left panel of Fig. 2. As it can be seen the RNN detected flares over 5σ with $\geq 80\%$ recall and precision. To check the reliability of the statistics, we injected flares into real, noisy light curves. The results follow the same trend drawn by the recovered manually-injected flares (orange curve in Fig. 2). The points that were flagged as part of a flare are highlighted in red in Fig. 1.

To test the performance in case of other light curves characterized by different noise statistics and instrumental effects, we predicted the location of flares (without re-training the network) in TESS light curves as well. While the recall shows similar results to the Kepler case, the precision is much worse (Fig. 2). After plotting and checking the false positive events, it turned out that a large number of flares are missing from the human-made input catalog we used to test the RNN. Three examples can be seen in the right panel of Fig. 2. As an RNN flags each point in a light curve, opposite to other, convolution-based networks, where each (pre-selected) flare is clipped from the whole data set and represented by an "image", the completeness of the training sample may have a significant impact on the performance of the networks, especially on the precision, leading to distorted results.

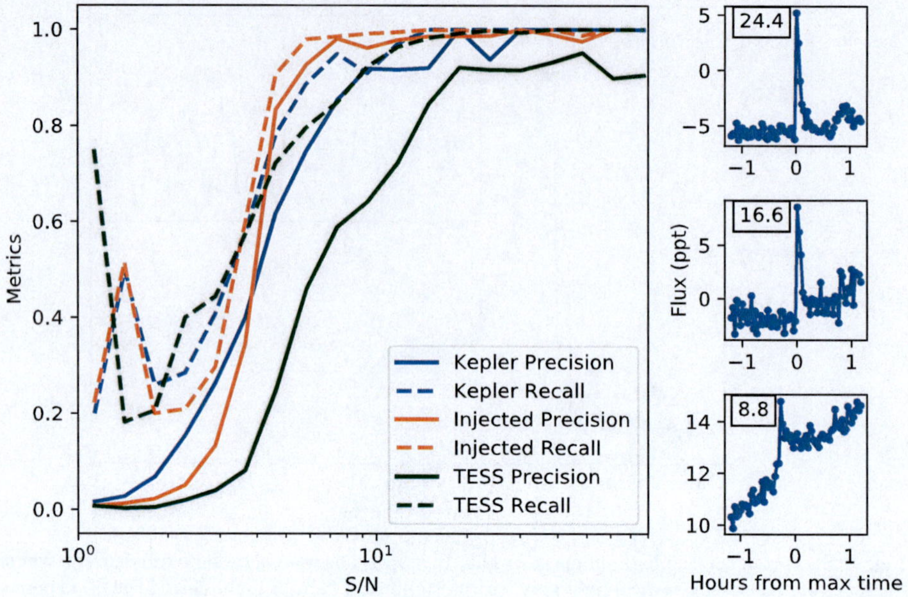

Fig. 2 Left: the precision/recall of the RNN as a function of flare signal-to-noise ratio. Blue curves are the Kepler observation, orange curves are the injected events, and green curves are the TESS observations. Right: false positive detections: missing flares from the TESS catalog, which were successfully identified by the RNN

3 Searching for Flares with `flatwrm2`

We have written a code that is intended to be used by the astronomical community to identify and characterize flares in Kepler and TESS space-based observations. The usage of `flatwrm2` and the necessary trained weight files can be found on our GitHub page.[2] Besides the labeling of flares in the given light curve, the code is also able to fit a model to each event to immediately return scientifically useful parameters.

4 Conclusions

We demonstrated that recurrent neural networks are suitable for searching for flares in evenly sampled, continuous spaced-based photometric observations.

[2] https://github.com/vidakris/flatwrm2.

Acknowledgments This project has been supported by the Lendület Program of the Hungarian Academy of Sciences, project No. LP2018-7, the NKFI KH-130526, NKFI K-131508 and 2019-2.1.11-TÉT-2019-00056 grants. On behalf of *"Analysis of space-borne photometric data"* project we thank for the usage of ELKH Cloud (https://science-cloud.hu) that helped us achieving the results published in this paper. Authors acknowledge the financial support of the Austrian-Hungarian Action Foundation (95 öu3, 98öu5, 101öu13).

References

1. Hardegree-Ullman, K.K., Cushing, M.C., Muirhead, P.S., Christiansen, J.L.: Kepler planet occurrence rates for mid-type M dwarfs as a function of spectral type. Astron. J. **158**, 75 (2019)
2. Jeffers, S.V., et al.: CARMENES input catalogue of M dwarfs. III. Rotation and activity from high-resolution spectroscopic observations. Astron. Astrophys. **614**, 76 (2018)
3. Günther, M.N., et al.: Stellar flares from the first TESS data release: exploring a new sample of M dwarfs. Astrophys. J. **159**, 60 (2020)
4. Khodachenko, M.L., et al.: Coronal Mass Ejection (CME) activity of low mass M stars as an important factor for the habitability of terrestrial exoplanets. I. CME impact on expected magnetospheres of earth-like exoplanets in close-in habitable zones. Astrobiology **7**, 167 (2007)
5. Yelle, R., Lammer, H., Ip, W.-H.: Aeronomy of extra-solar giant planets. Space Sci. Rev. **139**, 437 (2008)
6. Ranjan, S., Wordsworth, R., Sasselov, D.D.: The surface UV environment on planets orbiting M dwarfs: implications for prebiotic chemistry and the need for experimental follow-up. Astrophys. J. **843**, 110 (2017)
7. Seli, B., Vida, K., Moór, A., Pál, A., Oláh, K.: Activity of TRAPPIST-1 analog stars observed with TESS. Astron. Astrophys. **650**, 138 (2021)
8. Vida K., Roettenbacher, R.M.: Finding flares in Kepler data using machine-learning tools. Astron. Astrophys. **616**, A163 (2018)
9. Vida, K., Bódi, A., Szklenár, T., et al.: Finding flares in Kepler and TESS data with recurrent deep neural networks. Astron. Astrophys. **652**, A107 (2021)

Planetary Markers in Stellar Spectra: Jupiter-Host Star Classification

Miguel Andrea Zammit and Kristian Zarb Adami

Abstract Decades of observational and theoretical research have explored the relationship between a star's chemical, physical and galactic properties with the presence of orbiting planetary companions. Certain sources suggest that observed correlations are indicators of the environment of the system's protoplanetary disc, and subsequently its proclivity to facilitate planetary formation. This project aims to develop an ML classifier that uses spectral data of labelled target stars, to learn to model subtle discriminating markers and predict a binary class (Jupiter host or non-host) for every instance. We employ raw high-resolution stellar spectra as inputs, in order to preserve any inherent information within the spectrum, fed into a stacked ensemble architecture of several convolutional neural networks. Every CNN model is assigned a particular spectral range to collectively cover the entire spectrum, the results of which were then fed into a meta learner to aggregate their votes. The model achieves 96.14% test set accuracy, with 91.83% F_1 and 98.29% ROC-AUC scores.

1 Introduction

Several bodies of work in the literature have explored, and tended to show, intrinsic chemical differences between the stellar atmospheres of gas-giant host stars and those of comparison stars, with suggested links drawn to primordial abundances facilitating core accretion gas-giant formation (e.g. [1–3]).

Therefore, if there lies an intrinsic chemical discrepancy between gas-giant host stars and non-hosts, then such a pattern within the data could be tentatively used to predict which stars are likelier to host Jovian planetary companions, simply by analysing a star's spectroscopic data. Hence, we employ the power of computer vision in order to classify a star as a 'host' or 'non-host' of a gas-giant using high-

M. A. Zammit (✉) · K. Zarb Adami
Institute of Space Sciences and Astronomy, University of Malta, Msida, Malta
e-mail: mzamm33@um.edu.mt; kristian.zarb-adami@um.edu.mt

F. Bufano et al. (eds.), *Machine Learning for Astrophysics*, Astrophysics and Space Science Proceedings 60, https://doi.org/10.1007/978-3-031-34167-0_22

quality spectral data as input. It should be noted at this point that we aim to present models using abundance data in future work (similar in concept to the work found in [4]), to then cross-examine the performance of both methods. Here we apply a stacked ensemble CNN design for our model, and curate a dataset of 5417 high-resolution spectra from two ESO facilities (ESO-HARPS and ESO-FEROS).

2 Data Selection, Preparation and System Design

Prior to any model design and selection, it was vital that an extensive dataset was curated, cleaned, prepared and confidently labelled. We first started curating a list of stellar targets using the Hypatia Catalog [5], a spectroscopic abundance data catalog. It was ensured that each of the selected target systems could be confidently labelled and observed. A query was then ran on the ESO Science Archive Facility for publicly available high-resolution spectra observed by two instruments: the High Accuracy Radial Velocity Planet Searcher (HARPS) and the Fibre-fed, Extended Range, Échelle Spectrograph (FEROS), both at the ESO La Silla facility, selecting a maximum 25 spectra per target. As the selected spectra needed to have relatively high SNR ratios, only those whose SNR lies within the 50th percentile were eligible for selection. The final dataset's SNR distribution has a median of 165.6, and the percentage of instances with an SNR greater than 100 is 87.75%.

Training labels were assigned using a binary class designation on whether a star is a gas-giant host (1) or not (0), through consultation of planetary class designations in the ExoKyoto[1] database and the NASA Exoplanet Archive.[2] The final dataset comprised 5417 instances of 434 target stars and was sub-divided into 4 subsets: two training sets—one for the first-level classifiers (56%) and one for the meta learner (19%), a validation set to employ during training for hyperparameter tuning (9%), and a final holdout test set (16%). With the dataset curated, the spectra were truncated to ~3781–6912Å, and spectrally binned to a final resolution of 104,350 data points. We then normalised with respect to their spectral continuum, generated using the specutils[3] Python library. Each spectrum was subsequently separated into 25 disjoint subsets, each with 4174 data points to be fed to its assigned first-level classifier.

As stressed in the previous sections, the need for high-resolution spectra over considerable ranges of the visible regime posed an initial problem in applying CNN architectures. To avoid invoking the curse of dimensionality, we employ a stacked ensemble architecture with the first-level composed of individual, moderately deep CNN classifiers. Each of these voters is assigned a disjoint bin of the spectral axis, and outputs a probability score based on the spectral information within said bin.

[1] http://www.exoplanetkyoto.org/.

[2] https://exoplanetarchive.ipac.caltech.edu/.

[3] https://specutils.readthedocs.io/en/stable/index.html.

The CNN design for this stage of the project was chosen to be relatively shallow and linear, with 2 convolutional blocks followed by a number of simple dense layers for binary classification. The number of layers in each block, the size of the convolutional kernel and max-pooling bin, as well as the number of neurons of the first dense layer were set to be adjustable hyperparameters. This allowed us to tune each CNN to its respective spectral bin through a randomised cross-validation search, such that the hyperparameter space could be explored. Once the CNNs were all tuned and trained, the meta learner in the second level of the ensemble architecture was trained using the second training set. The meta learner is in essence tasked with receiving the probability scores of all CNN voters to aggregate and output as a final class label.

3 Results

With the model built and the dataset fully prepared, the first-level classifiers were trained on the first training set, with the validation set used to evaluate the model's generalisation accuracy over the course of the training epochs. A general trend that was prevalent during training was the models' more erratic performance on the validation set than on the training set, implying a certain degree of overfitting. This should be expected to manifest itself in a drop in performance of the meta learner during training on the second training set. Since the first-level classifiers do not approach their optimal transformation function in terms of generalisation, then the resulting inferred predicted scores on the second training set, which will be used as inputs to the meta learner, will hinder its ability to optimally generalise itself. An immediate solution to minimize overfitting in future work would be to include regularisation techniques into the architecture, as we recommend in Sect. 4.

Once the first-level voters were trained, they were used to infer predicted scores on the second training set, thus preparing the input dataset used to in-turn train the meta learner model. The performance metrics for the final model on the second training set can be found in Table 1. As expected, the performance of the final, trained model on the second training set suffers a slight dip, symptomatic of the first-level classifiers overfitting on the first training set. The model does however demonstrate strong generalisation on the two holdout sets, naturally slightly favouring the validation set as it was used to tune all the model hyperparameters. The

Table 1 Performance metrics of the meta learner

Dataset	Accuracy	Precision	Recall	F_1 score	ROC-AUC
Training set 1 (%)	99.31	98.71	98.29	98.50	99.82
Training set 2 (%)	94.98	90.67	87.18	88.89	97.92
Validation set (%)	97.05	92.79	94.50	93.64	98.01
Test set (%)	96.14	89.67	94.09	91.83	98.29

test set's ROC curve does however score slightly higher in its AUC than that of the validation curve, further suggesting clear, stable and relatively strong generalisation.

4 Conclusion and Future Work

In this body of work, we curate a labelled dataset of 5417 high-resolution stellar spectra of gas-giant hosts and comparison stars to train and test a stacked ensemble of CNN classifiers, to discriminate between host and non-host based solely on stellar chemistry. Our results show that the first-level classifiers have a general tendency to overfit the data, yet still exhibit strong collective generalisation. We aim to further extensively tackle the problem by first exploring techniques on how to minimize the extent of the model's overfitting to the training set. To do so, we aim to apply regularisation techniques and transfer learning. A second approach in using elemental abundance data as input features will also be implemented, and cross-examination of this with the current spectral input design will be conducted to assess which would yield greater learning and generalisation.

Acknowledgments This work is based on observations collected at the European Organisation for Astronomical Research in the Southern Hemisphere under ESO programmes.[4] This research has also made use of the NASA Exoplanet Archive, which is operated by the California Institute of Technology, under contract with the National Aeronautics and Space Administration under the Exoplanet Exploration Program. We would further like to acknowledge the use of the ExoKyoto Exoplanet Archive, developed and maintained by Kyoto University. The research shown here acknowledges the use of the Hypatia Catalog Database, an online compilation of stellar abundance data as described in [5], which was supported by NASA's Nexus for Exoplanet System Science (NExSS) research coordination network and the Vanderbilt Initiative in Data-Intensive Astrophysics (VIDA).

References

1. Gonzalez, G.: The stellar metallicity—giant planet connection. Month. Not. R. Astron. Soc. **285**(2), 403–412 (1997)
2. Reid, I.N.: On the nature of stars with planets. Publ. Astron. Soc. Pac. **114**(793), 306 (2002)
3. Thiabaud, A., Marboeuf, U., Alibert, Y., Leya, I., Mezger, K.: Elemental ratios in stars vs planets. Astron. Astrophys. **580**, A30 (2015)
4. Hinkel, N.R., Unterborn, C., Kane, S.R., Somers, G., Galvez, R.: A recommendation algorithm to predict giant exoplanet host stars using stellar elemental abundances. Astrophys. J. **880**(1), 49 (2019)
5. Hinkel, N.R., Timmes, F.X., Young, P.A., Pagano, M.D., Turnbull, M.C.: Stellar abundances in the solar neighborhood: the Hypatia Catalog. Astron. J. **148**(3), 54 (2014)

[4] The list of ESO programmes can be found in https://github.com/miguel-zammit-uom/jupiter-host-classification.

Using Convolutional Neural Networks to Detect and Confirm Exoplanets

Amelia M. Yu

Abstract In this research, I developed a Python software program with a convolutional neural network (CNN), a deep learning (DL) approach, mainly from TensorFlow to detect exoplanets through their transit signals in the National Aeronautics and Space Administration (NASA) Kepler space telescope data. My program first normalizes light curves, then creates and trains the deep learning model, tests and evaluates the model with sample data, folds the light curves of real data to intensify the transit signals, and subsequently applies the DL model to search for exoplanet transits. With this program and method, I detected four new exoplanets, as well as hundreds of confirmed and candidate exoplanets. All of the findings indicate that deep learning is an effective method that can be used to detect exoplanets in astronomy big data.

1 Introduction

In this project, I developed a Python software program with a convolutional neural network (CNN), a deep learning (DL) approach, mainly from TensorFlow to detect exoplanets through their transit signals in the National Aeronautics and Space Administration (NASA) Kepler space telescope data. The objective of this research is to develop an effective method using convolutional neural networks to find exoplanets among astronomical data, such as the Kepler data. In this research, my program first normalizes light curves, then creates and trains the deep learning model, tests and evaluates the model with sample data, folds the light curves of real data to intensify the transit signals, and subsequently applies the DL model to find exoplanets. With this program, I detected four new exoplanets, as well as hundreds of confirmed and candidate exoplanets. These four newly detected exoplanets are not listed in the Kepler Object of Interest (KOI) list. Among them,

A. M. Yu (✉)
Harvard University, Cambridge, MA, USA
e-mail: ameliayu@college.harvard.edu

F. Bufano et al. (eds.), *Machine Learning for Astrophysics*, Astrophysics and Space Science Proceedings 60, https://doi.org/10.1007/978-3-031-34167-0_23

two are ultra-short period (USP) exoplanets with orbital periods shorter than one day, and a possible Jupiter-like exoplanet. USP exoplanets are important subjects of astrophysics research because in order to orbit at such close distances to their stars, these USP exoplanets demonstrate special physical patterns, such as tidal interactions and spin evolution, and also challenge some astrophysical explanations. In addition, transit signals of exoplanets with extremely long periods are difficult to find in the Kepler data because the Kepler mission lasted for only 9.6 years and observes each star for a selected period of time. Thus, there are many more KOIs with shorter periods than those with longer periods in the NASA database. All of the findings indicate that deep learning is an effective method that can be used to detect exoplanets in astronomy big data, see e.g. [1, 2].

2 Data and Methods

The data used in this study have been retrieved from the public NASA Kepler database.[1] In this project, I used the light curves of 15,000 stars. The exoplanet search method is based on the transit method. The software has been developed using open source Python code and the TensorFlow framework, to implement an 8-layer convolutional neural network (CNN) to search for exoplanets. The program first creates a simulated artificial dataset with transit-like features and the trapezoid method, where varying parameters representing period, duration, depth, and ratio were used to create inverted parallel trapezoids that resembled transit signals. 100,000 light curves were used for training and 2000 light curves were used during validation testing. Second, the program normalized the light curves retrieved from the Kepler database. Third, the model has been trained with the Adam optimization algorithm to optimize the model parameters. Via training of 14 epochs, the model accuracy reached 0.94/1.00, and the loss was reduced to 0.09/1.00. Fourth, this program folded the transit light curves to intensify the transit signals. Finally, the code allows to search for transits in the light curve folds to detect exoplanets.

3 Results

This program detected four possible new exoplanets (shown in Fig. 1) when the conference abstract was first submitted in March 2022, marking the first step to discovery of these exoplanets. The new exoplanets include two ultra-short period (USP) exoplanets orbiting Kepler Input Catalog (KIC) 5881688, approximately 1475 parsec away from us, and KIC 9595827, approximately 940 parsec away from us. In KIC 1717722's multiplanetary system, in which two other candidate

[1] https://archive.stsci.edu/missions-and-data/kepler.

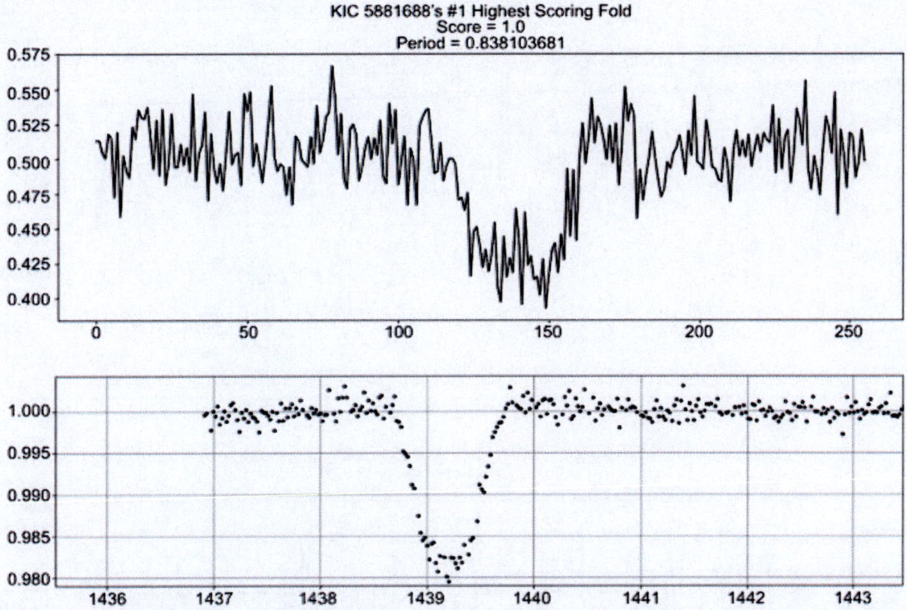

Fig. 1 The transit signals (U-shapes) for newly detected possible exoplanets. X-axis: time; Y-axis: flux

exoplanets with orbital periods shorter than 10 days have been reported, the software detected a possible Jupiter-like exoplanet. Figure 1 shows its transit. However, because only one light curve was detected, and we cannot exclude the possibility of error in the data, further work on this exoplanet is needed. The transit covered 2% of the star flux, which shows that planet size is comparable to that of Jupiter's. Out of the 1600 days that the Kepler telescope surveyed this star, the transit only appeared once. Thus, its orbital period is greater than 1600 days. Its radius is approximately 10.6 $R_{\oplus}$, which is close to Jupiter's radius of 11.2 $R_{\oplus}$. Meanwhile, Jupiter's planet-to-star radius ratio is 0.10, and this newly detected exoplanet's planet-to-star radius ratio is 0.13. The similar features of this planet indicate that the exoplanet detected in this research may be a Jupiter-like exoplanet. Since the light curve signal is so large, a ground-based telescope may be able to catch its transit signal for confirmation. The program also detected a new possible short-period exoplanet orbiting KIC 10975146. Table 1 lists the orbital periods, their scores out of 1.0, and their stellar parameters. This program also detected to again confirm hundreds of validated/confirmed and candidate exoplanets in the NASA exoplanet database.[2] Some examples are indicated in Table 2.

[2] https://exoplanetarchive.ipac.caltech.edu/.

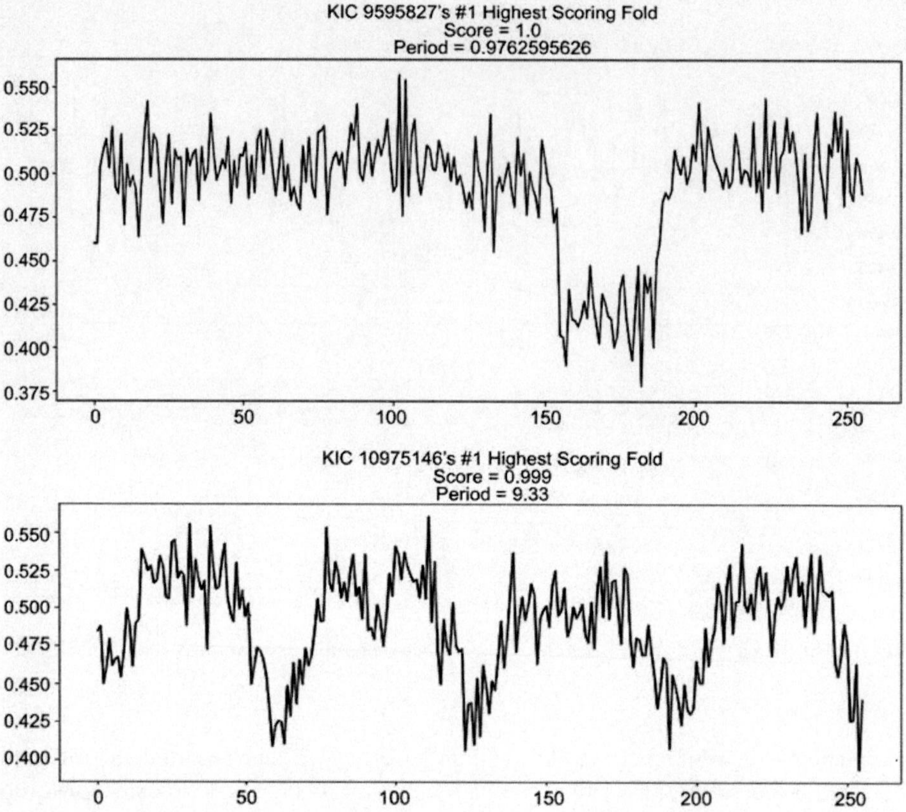

Fig. 1 (continued)

Table 1 Transit scores for the newly discovered exoplanets. Stellar data was retrieved from the open exoplanet catalogue and the NASA exoplanet archive

KIC	P (d)	Score	D (pc)	T_{eff}(K)	R ($R_\odot$)	M ($M_\odot$)
5881688	0.84	1.0	1475	5903	1.08	1.06
9595827	0.98	1.0	940	5543	0.89	0.92
1717722	>1600	–	592	4606	0.71	0.73
10975146	2.33	0.99	343	4653	0.76	0.71

4 Discussion and Conclusions

My program detected four new, previously undiscovered exoplanets. This is the first step to the discovery and confirmation of these exoplanets. Further work is needed to evaluate whether these detections came from false positives or signals of real exoplanets. Besides these newly detected exoplanets, this program also discerned hundreds of previously discovered, especially confirmed, exoplanets in the Kepler

Table 2 Comparison of orbital periods determined by the code and periods in literature for a sample of already confirmed and candidate exoplanets

KIC	Transit signals	P_{code} (d)	P_{lite} (d)	Status
1725016	Detected	7.41	7.41 (Kepler-748 b)	Confirmed
6964929	Detected	0.67	0.67 (Kepler-1340 b)	Confirmed
4144576	Detected	0.81	0.81 (Kepler-1139 b)	Confirmed
5942808	Detected	0.63	0.63	Candidate
4665571	Detected	0.77	0.77	Candidate
6525946	Detected	0.50	0.50	Candidate

data, demonstrating its capability to detect real exoplanets. To conclude, the results of this research demonstrate that deep learning, such as CNNs, is an effective or feasible method that can be used to discover exoplanets within astronomy big data, such as the Kepler data.

References

1. Pearson, K.A., Palafox, L., Griffith, C.A.: Searching for exoplanets using artificial intelligence. Month. Not. R. Astron. Soc. **474**(1), 478–491 (2018)
2. Zucker, S., Giryes, R.: Shallow transits—deep learning. I. Feasibility study of deep learning to detect periodic transits of exoplanets. Astron. J. **155**(4), 147 (2018)

Machine Learning Applied to X-Ray Spectra: Separating Stars from Active Galactic Nuclei

Pavan R. Hebbar and Craig O. Heinke

Abstract Modern X-ray telescopes have detected hundreds of thousands of X-ray sources. However, current methods to classify these sources using the X-ray data themselves suffer problems—detailed modelling of X-ray spectra for each source is very inefficient, while hardness ratios can be confusing for faint sources. In this paper, we probe the use of artificial neural networks to distinguish between line- and continuum-dominated sources and differentiate the young stars in the Chandra Orion Ultradeep Project (COUP) survey from the active galactic nuclei (AGN) in the Chandra Deep Field South (CDFS) survey. We find that our methods reach an accuracy of ~92% in classifying simulated spectra of moderate-brightness objects in typical exposures, but their performance decreases on the observed COUP and CDFS spectra (~85–90%), due in large part to the relatively high background of these long-exposure datasets. We also investigate the performance with changing properties of the spectra such as the net source counts, the contribution of background, the absorption column of the sources, the thermal temperatures of stars etc. We conclude that these methods are promising for application to large X-ray surveys.

1 Introduction

Over the last decades we have launched several powerful X-ray telescopes to observe high energy X-ray sources in the universe such as the *Chandra X-ray Observatory* (*Chandra*, for short). The *Chandra* Source Catalog (CSC) has ~300,000 unique sources (Evans et al. [1]). These include active galactic nuclei (AGN), supernova remnants (SNRs), active stars, and compact neutron star and black hole binaries.

P. R. Hebbar (✉) · C. O. Heinke
Department of Physics, CCIS 4-181, University of Alberta, Edmonton, AB, Canada
e-mail: hebbar@ualberta.ca; heinke@ualberta.ca

F. Bufano et al. (eds.), *Machine Learning for Astrophysics*, Astrophysics and Space Science Proceedings 60, https://doi.org/10.1007/978-3-031-34167-0_24

Common methods to distinguish these sources have limits—detailed X-ray spectroscopy of thousands of sources is tedious, and hardness ratios could fail to accurately categorize faint objects. These methods fail to employ the power of CCD detectors to identify emission lines in the X-ray spectra. Machine-learning (ML) tools that can differentiate line- and continuum-dominated X-ray spectra can classify the sources efficiently. Here, we demonstrate using artificial neural networks (ANNs) to distinguish stars in the Orion nebula from AGN in the Chandra Deep Field South.

2 Data Acquisition and Analysis

We use the energy spectra of X-ray sources in the *Chandra* Orion Ultradeep Project (COUP) survey (Getman et al. [2]), and the *Chandra* Deep Field South (CDFS) AGN spectral properties catalog (Tozzi et al. [3]) to generate the artificial spectra. The stars were generated using the models `tbabs*apec` and `tbabs*(apec+apec)` in XSPEC. In the case of AGN, 93% of them use *an absorbed power-law model* (`cflux*tbabs*ztbabs*pegpwrlw`), 3% of them use *an absorbed power-law with a soft component* (`tbabs*pegpwrlw + cflux*tbabs*ztbabs*pegpwrlw`), and 4% use *a Compton thick emission model* (`cflux*tbabs*pexrav`). We also add a redshifted Fe-K line to 50% of these spectra.

We use the flux of CDFS AGN in the 0.5–10 keV range to simulate stars so that both stars and AGN spectra have similar signal-to-noise ratios. Since the CDFS AGN are much fainter than typical sources detected by *Chandra*, we increase the flux of AGN by 10 times and simulate the AGN spectra for an exposure of 100 ks (i.e. we reduce the contribution of background by 10 while maintaining the source counts).

We incorporate a single-layer eight-node sequential artificial neural network with rectified linear unit (ReLU) activation and an L1 regularization, $\lambda = 0.001$ for the training. *We use the binary cross-entropy loss, adam method and 20 iterations for optimization.* We only consider 0.3–8.0 keV energy range to minimize the Poisson noise and background contribution. We use 80% of the data for training and 20% as the test set. We also perform a 10-fold cross-validation to avoid over-fitting.

3 Results and Discussion

We achieve a net classification accuracy of ~92% with our model (Fig. 1). Applying this trained model to the observed spectra gives an accuracy of ~91%. The difference in accuracy could be due to the reduced background in the simulated spectra. The best-fitting weights of our ANN model indicate that it can identify the Fe-L, Mg-K, Si-K and S-K emission lines in stars and differentiate them from

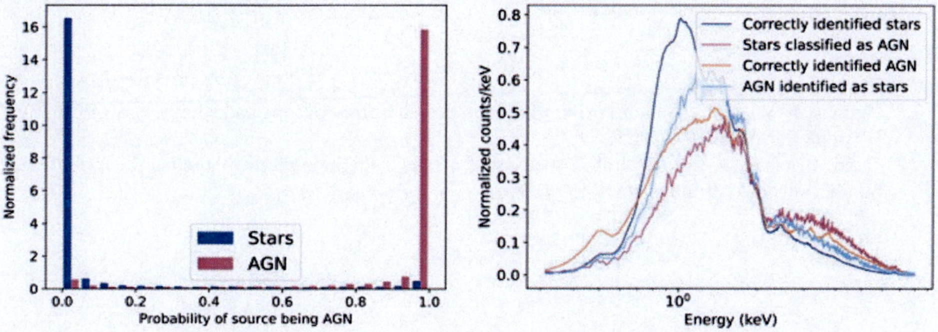

Fig. 1 *(Left)* output probabilities from the ANN of the source being AGN. Most of the stars and AGN are properly identified with high confidence. *(Right)* mean spectra of correctly and incorrectly identified stars and AGN. Stars identified as AGN do not have strong emission lines

the continuum spectra of AGN. We also analyze the variation in the performance of our model with the properties of the spectra. We notice that the accuracy of classification increases with increasing net counts and decreasing background-to-net count ratio (due to higher signal to noise ratio; our accuracy $\gtrsim$90% for net counts $\gtrsim$200 and background-to-net count ratio $\lesssim$0.05.) For $N_H > 10^{22}$, the emission lines in stars are absorbed and our model fails to identify them accurately. Extremely hot temperatures of stars ($>$ 2 keV) also lead to complete ionization of the elements, removing visible lines. Our model also tends to identify harder AGN, with power-law index $\Gamma < 2$, better. We also notice that AGN with strong Fe-K emission lines are identified more efficiently.

4 Conclusions

We showed that artificial neural networks could be efficient tools to distinguish line- vs. continuum-dominated spectra and thus classify X-ray sources. Our model works best for sources with net counts greater than 200, and background contribution less than 5%. Using the information from X-ray spectra along with multi-wavelength properties can allow us to achieve more robust classification of X-ray sources.

X-ray catalogs usually have several thousands of sources with only a few bright sources labelled. Thus, the use of clustering methods, projecting the data into lower-dimensions through auto-encoders etc., can allow us to approach the problem through semi-supervised learning. We plan to extend our techniques to other kinds of X-ray sources like neutron stars and SNRs, and to different X-ray catalogs.

References

1. Evans, I.N., et al.: The chandra source catalog. Astrophys. J. Suppl. Ser. **189**, 37 (2010)
2. Getman, K.V., et al.: Chandra orion ultradeep project: observations and source lists. Astrophys. J. Suppl. Ser. **160**, 319 (2005)
3. Tozzi, P., et al.: X-ray spectral properties of active galactic nuclei in the Chandra Deep Field South. Astron. Astrophys. **451**, 457 (2006)

Classification of System Variability Using a CNN

J. Magdolen and A. Dobrotka

Abstract The nature of detected variability (whether the frequency is variable or stable) is decisive in binary systems. Dobrotka and Ness (Month Not R Astron Soc 467:4865–4871, 2017) showed that a stable frequency with variable amplitude can mimic the presence of two close frequencies, which can lead to a false interpretation of the signal. Based on XMM-Newton and Chandra observations of binaries Cal 83 and KT Eri respectively, we performed multiple simulations of light curves with different types of variability. Both systems show the double-peak pattern and were previously discussed as variable in frequency (Ness et al., Astron Astrophys 578:A39 2015). Next, we created their Dynamic Power Spectra (DPS) and implemented a Convolutional Neural Network algorithm (CNN) in order to find some possible hidden features inside DPSs and thereby identify the origin of the variability.

1 Introduction

The Cal 83 and KT Eri are binary systems with 67s and 35s period respectively, containing a hot primary star—white dwarf (WD) and a cool companion—late main sequence star. Both systems were observed in so-called SSS (Super Soft Source) phase. The hydrogen-rich matter from the companion falls onto the WD surface through accretion. Based on the accretion rate, sustained nuclear burning or nova outburst is present.

By creating a light curve (LC) and then a periodogram, it is possible to determine the significance of certain frequencies as well as their character. Observation of both systems show a double-peak pattern in their power spectra, what can suggest two close frequencies or a variable frequency. However, in the paper [1] was demonstrated that a typical double-peak pattern in periodogram can be the result not

J. Magdolen (✉) · A. Dobrotka
Advanced Technologies Research Institute, Faculty of Materials Science and Technology in Trnava, Slovak University of Technology in Bratislava, Trnava, Slovakia
e-mail: jozef.magdolen@stuba.sk; andrej.dobrotka@stuba.sk

F. Bufano et al. (eds.), *Machine Learning for Astrophysics*, Astrophysics and Space Science Proceedings 60, https://doi.org/10.1007/978-3-031-34167-0_25

only of two close frequencies, but also of one frequency with a variable amplitude. Therefore, potential misinterpretation of a WD rotation as a pulsation on the WD surface is possible. By using CNN AlexNet algorithm, we tried to find out whether it is possible to distinguish between variability in amplitude and variability in frequency.

Since we focus on novae in SSS phase, there are not many observed similar sources in the archives. For that reason we simulated multiple LCs with different types of variability based on Cal 83 and KT Eri sources and created their DPSs.

2 Data Simulation and CNN

Taking *Chandra/HRC* observation for KT Eri (ObsID 12203) and *XMM-Newton/pn* observation for Cal 83 (ObsID 0506531701) respectively, we simulated synthetic LCs using a sinusoidal function $\psi = \phi + P_a \sin(2\pi t / P_p)$, where ϕ is a mean value. P_a and P_p are 13th degree polynomials describing variable amplitude a and/or variable period p. These were obtained by generating a distribution of Gaussian points around the a and p mean values with corresponding variances, where the a mean value was estimated by creating a folded LC for each system. Next, the points were fitted with a 25 degree polynomial, which is the function used as input for the mentioned sine function. If the modulated amplitude function P_a is less than 0, we assume $P_a = 0$ because there is no meaning to negative amplitude. Such case represents situation when the variability disappears for a while. Next, a Poisson noise was added.

For 4 models of variability (namely constant amplitude with constant frequency, constant period with variable amplitude, variable period with constant amplitude, and variable both amplitude and period) we simulated 5000 LCs per each system. Next, we broke each LC into segments and created Lomb-Scargle (LS) periodograms for each segment. By producing the DPS by adding all periodograms into one color coded figure, the evolution of power in time is illustrated. DPSs created in such a way were used as the input to the CNN.

3 Results and Discussion

We used the AlexNet CNN architecture to determine if it is able to classify the type of variability by which the observed systems are modulated. The CNN trained on simulated Cal 83 data reports 92% accuracy and the system itself was classified as variable in amplitude and frequency, unlike in [2], where it was classified as variable in frequency only. Contrariwise, the accuracy of the model trained on KT Eri data reached only 50%, what is too low to reliably classify any system. The CNN accuracy and the systems classification are shown in Table 1.

Table 1 CNN accuracy and systems classification

System	Const a. Const. f. (%)	Var a. Const. f. (%)	Const a. Var. f. (%)	Var a. Var. f. (%)	Accuracy (%)
Cal 83	0.00	0.03	0.00	99.97	92.1
KT Eri	0.01	0.01	51.07	48.91	49.8

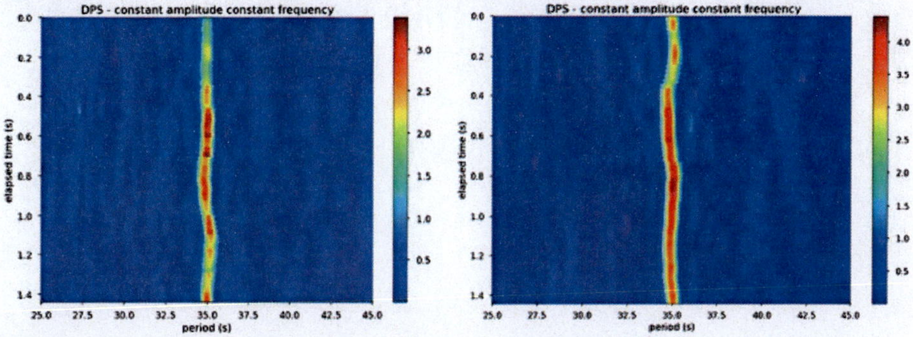

Fig. 1 The example of DPSs simulated with constant amplitude and constant frequency, but the variability in signal is visible

To create a DPS requires to compute periodograms and subsequently the power must be color coded. Such processing can lead to worse classification, since the color can create some artefacts, as can be seen in the background of the DPSs in Fig. 1. Therefore, we will repeat the learning process with periodograms and afterwards with LCs as we want to find out, whether the accuracy of the CNN will improve.

We also found some DPSs similar to those simulated with variable frequency, but **constant frequency and constant amplitude** was used (see Fig. 1). Therefore, we will need to look deeper into data and also determine the statistical significance of such results. If the constant frequency with noise alone can mimic the variability in frequency, that will give us new insight on the variability not only in binary systems. In such case any apparent variability in examined system must be taken with caution.

Acknowledgments This work was supported by the Slovak grant VEGA 1/0408/20, and by the European Regional Development Fund, project No. ITMS2014+: 313011W085.

References

1. Dobrotka A., Ness J.-U.: Counter-evidence against multiple frequency nature of 0.75 mHz oscillation in V4743 Sgr. Month. Not. R. Astron. Soc. **467**, 4865–4871 (2017). https://doi.org/10.1093/mnras/stx442
2. Ness J.-U., Beardmore A.P., Osborne J.P., Kuulkers, E., Henze, M., Piro, A.L., Drake J.J., et al.: Short-period X-ray oscillations in super-soft novae and persistent super-soft sources. Astron. Astrophys. **578**, A39 (2015). https://doi.org/10.1051/0004-6361/201425178

Deep Learning Processing and Analysis of Mock Astrophysical Observations

Claudio Gheller, Matteo Angelinelli, Klaus Dolag, Nicoletta Sanvitale, and Franco Vazza

Abstract Deep Learning represents a promising, general purpose solution to process and analyse large, complex datasets in an automated and efficient fashion. We present the experience accomplished on radioastronomy and X-ray simulated astrophysical data, adopting two different approaches, Autoencoders and Fully Convolutional Networks. The former aims at denoising the data, the latter at detecting and identifying faint sources in noisy images. We give an overview of the main outcomes.

1 Introduction

In the upcoming decade, astronomers will have to face novel scientific and technological challenges. An avalanche of complex data will be delivered by current and upcoming telescopes. This will be difficult to manage with traditional approaches. Data will have to be stored in dedicated facilities, providing the necessary capacity at the highest performance. Corresponding data processing will have to be performed local to the data, exploiting available high performance computing resources. Data reduction and imaging software tools will have to be adapted, if not completely re-designed, in order to efficiently run at scale. Fully automated pipelines will be a

C. Gheller (✉)
IRA, Bologna, Italy
e-mail: claudio.gheller@inaf.it

M. Angelinelli
CINECA, Casalecchio di Reno, Italy

K. Dolag
Universitäts-Sternwarte München, München, Germany

N. Sanvitale
Institute for Theoretical Physics, University of Cologne, Cologne, Germany

F. Vazza
University of Bologna, Bologna, Italy

F. Bufano et al. (eds.), *Machine Learning for Astrophysics*, Astrophysics and Space Science Proceedings 60, https://doi.org/10.1007/978-3-031-34167-0_26

compelling requirement for effective software stacks as the richness and complexity of incoming data will inhibit human interaction and supervision.

In order to develop such data processing methodologies, we have explored the potential of Machine Learning, a branch of Artificial Intelligence already successfully used in astronomy. Among the various Machine Learning approaches, we have focused on Deep Learning, which provides outstanding performance for tasks relating to computer vision, text analysis, fragmentation, speech recognition (see [1, 2] and reference therein) among others. Deep Learning has become increasingly popular in the last decade thanks to two concurrent factors: the availability of enough computing power to cope with complex, multi-layered neural networks, and the availability of enough data to perform the training. We have adopted Deep Learning solutions with two different objectives: the denoising of radio astronomy images, that will be described in Sect. 2, and the detection of X-ray sources in noisy data, that will be discussed in Sect. 3.

2 Denoising Radio Interferometric Data

In the case of radio interferometry, the aperture synthesis approach is exploited to generate observations in the radio band. Aperture synthesis consists in the adoption of sophisticated data processing techniques in order to obtain sky images from the collected signals, estimating the energy flux density coming from a given region of the sky, at a specific frequency. Besides random noise (of, e.g., thermal or electronic origin), these observations are affected by the resolution of the instrument, by the geometric pattern characterising the observation process (sparse antennas rotating with the Earth) and by a number of other effects related to the propagation of radio signals through the Earth's atmosphere (e.g. ionospheric disturbances or Radio-Frequency Interference) which may contaminate the final product. For a more extensive discussion of these aspects, we refer to [2] and reference therein.

We have adopted a Deep Learning based methodology to remove such noise and artefacts from images resulting from the imaging and deconvolution processes on data mimicking LOFAR HBA observations. More specifically, we have adopted an *autoencoder* based approach. An autoencoder is a type of neural network which aims at learning an identity map for the input data through encoding and decoding data pairs:

$$\tilde{I} = D(E(I)), \tag{1}$$

where I and $\tilde{I}$ are the input and reconstructed images respectively, and E and D are the encoder and decoder transforms, which minimise the difference of the reconstructed and the input data:

$$D, E = \text{argmin}_{(\text{D,E})}(\text{MSE}(I - D(E(I))), \tag{2}$$

where MSE is the mean squared error. The architecture of the network consists in two convolutional plus two pooling layers for encoding, two convolutional with two unpooling layers for decoding and one dense layer made of 200/400 neurons. The 200 elements dense layer is to prefer for noisy data to avoid overfitting.

Realistic mock observations (Sky models), necessary for the training of the autoencoder, have been generated starting from light cones built from the results of cosmological simulations as described in [1] and [3]. A set of 1000 independent sky model images was generated by applying random rotations to each of the different redshift slices used to produce the lightcones. The nominal angular resolution of the images (pixel size) is 2 arcsec. The described procedure was used to generate 2000×2000 pixel images, both sky models and corresponding noisy images, sampling a field of view of $1.1^\circ \times 1.1^\circ$ and represent the dataset against which the autoencoder is trained.

Figure 1 shows an example of the denoising of an image where faint radio sources are present. In general, the autoencoder has proved to efficiently remove noise and artefacts preserving the properties of the regions of the sources with signal to noise ratio (S/N) of the order or lower than 1. However, the convolutional, multi-layer architecture of the encoder introduces numerical blurring which impacts in particular the highest peaks of the flux density distribution which result to be dimmed. Therefore the optimal adoption of the autoencoder seems to be for faint (i.e. at S/N $\sim$ 1 or below) sources and/or for performing data segmentation.

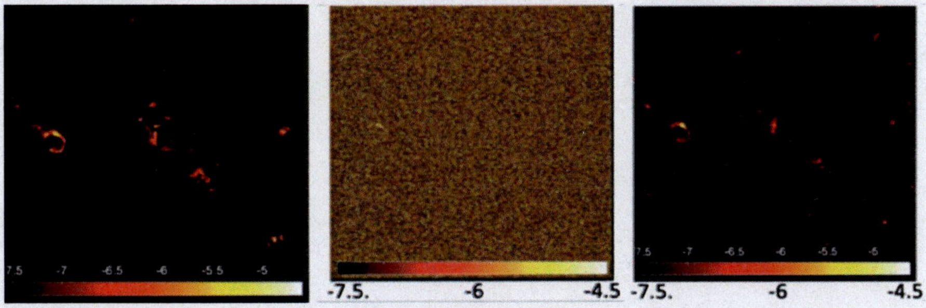

Fig. 1 Flux density maps of a sample sky model comparing the sky reference image (left), the noisy input image (center) and the autoencoder denoised image (right). Flux densities are in log(Jy/arcsec^2) scale

3 X-Ray Source Finder

In the case of X-ray observations, we have followed a similar procedure as that described in Sect. 2 to generate the mock images. In this case, however, we started from the Magneticum simulation [4] and we emulated Athena X-ray observatory data.[1]

The objective was to identify sources in a noisy background. We have adopted a *fully convolutional network* approach, that proved to be highly effective for the tasks related to image segmentation (e.g. [5]). It is interesting to notice that our Deep Learning model has been developed to perform the segmentation of data from a completely different scientific application, i.e. photoelastic images from experiments of granular materials [6] (see Fig. 2). However, it proved to be effective also for our astronomical data, highlighting the flexibility and the strength of Machine Learning.

The fully convolutional network consists of a contracting path and an expansive path. The contracting path is composed of an input layer, loading the input images, a downsampling convolutional network (composed by convolutional and pooling layers) which ends with a final level composed by convolutional layers. The expansive path starts from this deepest level, followed by an upsampling

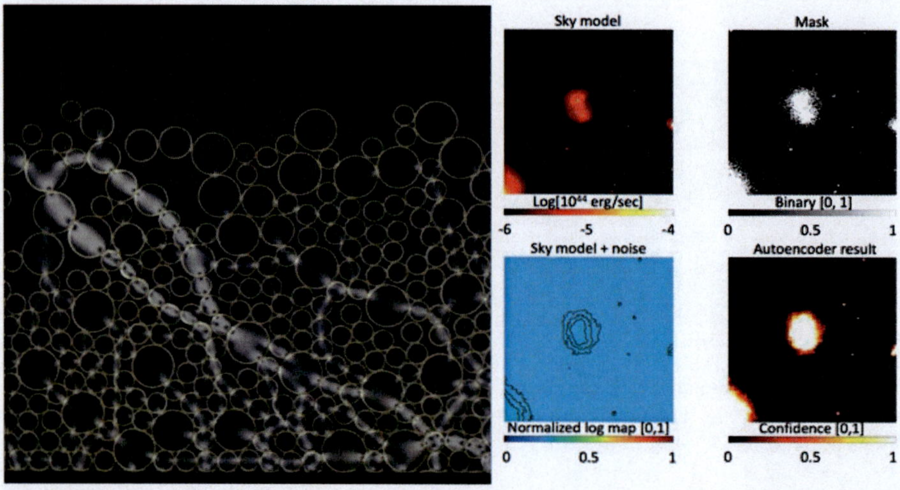

Fig. 2 Results of the application of the fully convolutional network for the segmentation of experimental granular flow data (left) and X-ray data (right). For X-ray data, the top left panel is the simulated X-ray Flux, the top right panel is the mask used for the training, the bottom left panel is the noisy image with probability contours superimposed and the bottom right panel is the probability distribution resulting from the network

[1] https://www.the-athena-x-ray-observatory.eu/.

convolutional network, specular to the downsampling one. At each upsampling level, the feature maps are summed to the corresponding feature maps from the contracting path, in order to supply the spatial information to precisely localise the identified features. The final output layer returns the results. In our case, the results are represented by the confidence maps, of the same size of the input ones, expressing the probability that a given pixel represents (or is part of) a source, as shown in Fig. 2.

4 Conclusions

Incoming astronomical observational facilities are going to deliver huge amounts of complex data. This data represents an ideal test case for novel Deep Learning approaches, that will effectively exploit increasingly capable computing resources. We have studied the usage of autoencoders to remove noise and artefacts from radio interferometric data, and the usage of fully convolutional networks for image segmentation. In both cases, the Deep Learning networks proved to be flexible and effective in their tasks, although further investigation is required to extend their adoption to observational data.

References

1. Gheller, C., Vazza, F., Bonafede, A.: Month. Not. R. Astron. Soc. **480**, 3749 (2018). https://doi.org/10.1093/mnras/sty2102
2. Gheller, C., Vazza, F.: Month. Not. R. Astron. Soc. **509**, 990 (2022). https://doi.org/10.1093/mnras/stab3044
3. Vazza, F., Paoletti, D., Banfi, S., et al.: Month. Not. R. Astron. Soc. **500**, 5350 (2021). https://doi.org/10.1093/mnras/staa3532
4. Hirschmann, M., Dolag, K., Saro, A., et al.: Month. Not. R. Astron. Soc. **442**, 2304 (2014). https://doi.org/10.1093/mnras/stu1023
5. Ronneberger, O., Fischer, P., Brox, T.: arXiv:1505.04597 (2015)
6. Sanvitale, N., Gheller, C., Bowman, E.: Granular Matter. **24**, 65 (2022). https://doi.org/10.1007/s10035-022-01222-w

Deep Neural Networks for Source Detection in Radio Astronomical Maps

Renato Sortino, Daniel Magro, Eva Sciacca, Simone Riggi, and Giuseppe Fiameni

Abstract Source finding is one of the most challenging tasks in upcoming radio continuum surveys with SKA precursors, such as the Evolutionary Map of the Universe (EMU) survey of the Australian SKA Pathfinder (ASKAP) telescope. In this work, we explore the application of several deep learning methods, based both on classical CNNs and on more recent transformer architectures to perform object detection. All the models were trained using ASKAP data, taken during the Early Science phase, and previous radio survey data. Results in terms of Precision, Recall, and F1-Score are promising. We evaluate several training strategies and model architectures to determine which one fits the case study best. Finally, we exploited High Performance Computing to run our experiments, to evaluate the improvement of a distributed training process, running on multiple computing nodes, compared to a single-node, single-process approach.

R. Sortino (✉)
Universitá degli studi di Catania, Catania, Italy
e-mail: renato.sortino@inaf.it

D. Magro
Institute of Space Sciences and Astronomy, University of Malta, Msida, Malta

INAF, Osservatorio Astrofisico di Catania, Catania, Italy
e-mail: daniel.magro.15@um.edu.mt

E. Sciacca · S. Riggi
INAF, Osservatorio Astrofisico di Catania, Catania, Italy
e-mail: eva.sciacca@inaf.it

G. Fiameni
NVIDIA AI Technology Center, Milan, Italy
e-mail: gfiameni@nvidia.com

F. Bufano et al. (eds.), *Machine Learning for Astrophysics*, Astrophysics and Space Science Proceedings 60, https://doi.org/10.1007/978-3-031-34167-0_27

1 Introduction

In recent years, the technological advancement in astronomy and astrophysics has marked the need for tools and innovative techniques for processing the huge amount of data and images captured from the different instruments for radio astronomy observations [1, 2]. In particular, in the context of radio-astronomical surveys, the task of object detection and instance segmentation is crucial for extracting information from images to support the astrophysics research to catalogue and identify the contained objects [3].

Previously, source detection consisted in visual inspection by trained astronomers, which is a reliable, but not scalable, technique, considering the volume of data produced by today's telescopes. Algorithmic techniques to automate the source finding task represent the next step, and include a variety of methods based on thresholding and peak detection such as AEGEAN [4] and PyBDSF [5].

Deep learning methods are capable of generalizing to unseen data. ConvoSource [6] uses a lightweight CNN to generate a binary mask; DeepSource [7] enhances performance by boosting the signal-to-noise ratio of input images; CLARAN [8] performs transfer learning from the ImageNet [9] dataset on a Faster R-CNN architecture; Astro R-CNN [10] applies Mask R-CNN [11] to detect sources on a simulated dataset.

These works, while representing pivotal contributions to the scientific community, lack of the following aspects: they have been trained on simulated data, assuming a particular kind of telescope, which limits the generalization capabilities.

In this work we evaluate and compare the performance of multiple object detection models based on deep learning, applied on astronomical radio images collected from different datasets, with the task of source detection and classification, in order to evaluate the main differences between these approaches.

2 Dataset: Galactic Radio Images

To train and validate the models we made use of a dataset containing 10,952 grayscale image cutouts extracted from different Galactic radio-astronomical survey images. The whole dataset consists of a total of 36,398 objects, then split into 3 subsets: training (70%), validation (10%) and test (20%).

Image cutouts (Fig. 1) (single-channel, 132×132 pixels, FITS format) were extracted from the reference data using the CAESAR tool [12]. The RGZ dataset already provides source cutouts in the same format, as well as bounding boxes for extended source objects. Images are preprocessed by setting 'NaN' pixels to the image minimum value and applying Z-Scale [13] normalization. During training, we employ image augmentation to prevent the model from overfitting. A random number of augmentations (0 to 2) is applied to each image before being fed to the model during training.

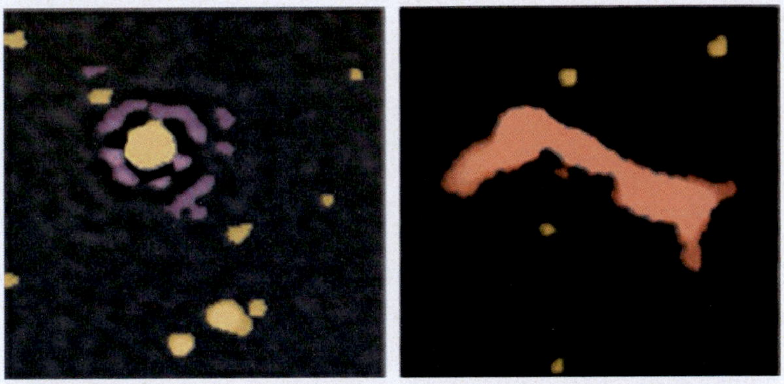

Fig. 1 Image and annotation sample pairs used in training. (Left) compact sources (in yellow) and spurious sources (in purple). (Right) extended source sample (in red)

3 Models

We investigated several object detection models, with the goal of identifying and classifying sources and other celestial objects. We mainly explored architectures based on CNNs (ASGARD) and on Transformers [14] (DETR, YOLOS).

- **ASGARD** [15] is a Mask R-CNN based model which was specifically adapted for working on radio-astronomical images and employs the following components: Feature Pyramid Network (FPN) to extract features from the image at multiple scales; Region Proposal Network (RPN), used to determine regions likely to contain an object of interest; ROI Pooling to resize the output of the RPN to a fixed size; and, Fully Convolutional Network (FCN), that generates a per-pixel mask for the proposed regions. ASGARD applies post-processing steps to its predictions and contains functionality for the processing of very large images, and not only smaller-sized cutouts.
- **DETR** [16] consists of 3 main modules: (1) a CNN backbone for feature extraction; (2) an encoder-decoder transformer, which computes self- and cross-attention on the features extracted by the backbone and produces a fixed set of predictions; (3) a feedforward network to predict the location and class of each output in the set. This method is capable of accurately predicting complex and overlapping objects, by using attention maps, but comes with non-trivial training, and some degree of error on smaller objects.
- **YOLOS** makes use of a fully transformer-based architecture similar to ViT [17], while DETR still relies on a convolutional backbone. YOLOS concatenates the patches of the input image with a series of learnable tokens, and computes self attention on this sequence, by means of a transformer encoder. With respect to DETR, YOLOS does not make use of the transformer decoder module, yielding fewer parameters and thus less computational burden.

Table 1 Detection metrics. BS stands for batch size. Best results highlighted in bold

Model	BS	Training	Reliability	Completeness	F1 Score	mAP@50
ASGARD	32	From scratch	0.5199	**0.7947**	0.6286	70.23
DETR	2	From scratch	0.1531	0.1842	0.1672	13.41
	2	Finetuned	**0.7642**	0.7683	**0.7662**	**78.95**
YOLOS	2	From scratch	0.4812	0.2594	0.3371	26.43
	2	Finetuned	0.5804	0.7550	0.6563	76.29

4 Performance Analysis

In Table 1, we report the results for the described models. DETR achieves the best performance in terms of reliability, F1-score, and mAP, while ASGARD shows the highest completeness.

ASGARD was trained from scratch, while DETR and YOLOS have been finetuned starting from weights pretrained on the COCO [18] dataset. This choice stems from the difficulty of reaching convergence with transformers when trained from scratch.

DETR yields the best results, but this comes at a high computational cost.

5 Conclusions

With the advent of the Square Kilometre Array, automated techniques are needed to efficiently analyse the data deluge expected to be produced by such new generation astronomical facilities.

In this work we evaluated the performance of some of the main deep learning approaches when applied to the source detection task in the radio-astronomy field, providing results for methods that, to the best of our knowledge, have not yet been applied to this task.

Acknowledgments The research has received funding from the EC Horizon 2020 grant agreement No. 863448 (NEANIAS) and from the project MOSAICo, co-funded by the EU.

References

1. Magro, D., Zarb Adami, K., DeMarco, A., Riggi, S., Sciacca, E.: A comparative study of convolutional neural networks for the detection of strong gravitational lensing. Month. Not. R. Astron. Soc. **505**(4), 6155–6165 (2021). https://doi.org/10.1093/mnras/stab1635
2. Ralph, N.O., Norris, R.P., Fang, G., Park, L.A.F., Galvin, T.J., Alger, M.J., Andernach, H., Lintott, C., Rudnick, L., Shabala, S., Wong, O.I.: Radio galaxy zoo: unsupervised clustering of convolutionally auto-encoded radio-astronomical images. Publ. Astron. Soc. Pac. **131**(1004), 108011 (2019). https://doi.org/10.1088/1538-3873/ab213d

3. Karypidou, S., Georgousis, I., Papakostas, G.A.: Computer vision for astronomical image analysis. In: 2021 IEEE International Conference on Progress in Informatics and Computing (PIC), pp. 94–101 (2021). https://doi.org/10.1109/PIC53636.2021.9687023
4. Hancock, P.J., Trott, C.M., Hurley-Walker, N.: Source finding in the era of the ska (precursors): Aegean 2.0. Publ. Astron. Soc. Australia **35**, e011 (2018). https://doi.org/10.1017/pasa.2018.3
5. Mohan, N.R., Rafferty, D.A.: PyBDSF: python blob detection and source finder. Astrophysics Source Code Library pp. ascl–1502 (2015)
6. Lukic, V., de Gasperin, F., Brüggen, M.: Convosource: radio-astronomical source-finding with convolutional neural networks. Galaxies **8**(1), 3 (2020)
7. Vafaei Sadr, A., Vos, E.E., Bassett, B.A., Hosenie, Z., Oozeer, N., Lochner, M.: Deepsource: point source detection using deep learning. Month. Not. R. Astron. Soc. **484**(2), 2793–2806 (2019)
8. Wu, C., Wong, O.I., Rudnick, L., Shabala, S.S., Alger, M.J., Banfield, J.K., Ong, C.S., White, S.V., Garon, A.F., Norris, R.P., et al.: Radio galaxy zoo: claran–a deep learning classifier for radio morphologies. Month. Not. R. Astron. Soc. **482**(1), 1211–1230 (2019)
9. Russakovsky, O., Deng, J., Su, H., Krause, J., Satheesh, S., Ma, S., Huang, Z., Karpathy, A., Khosla, A., Bernstein, M., et al.: Imagenet large scale visual recognition challenge. Int. J. Comput. Vis. **115**(3), 211–252 (2015)
10. Burke, C.J., Aleo, P.D., Chen, Y.C., Liu, X., Peterson, J.R., Sembroski, G.H., Lin, J.Y.Y.: Deblending and classifying astronomical sources with mask R-CNN deep learning. Month. Not. R. Astron. Soc. **490**(3), 3952–3965 (2019)
11. He, K., Gkioxari, G., Dollár, P., Girshick, R.: Mask R-CNN. In: Proceedings of the IEEE International Conference on Computer Vision, pp. 2961–2969 (2017)
12. Riggi, S., Vitello, F., Becciani, U., Buemi, C., Bufano, F., Calanducci, A., Cavallaro, F., Costa, A., Ingallinera, A., Leto, P., et al.: Caesar source finder: recent developments and testing. Publ. Astron. Soc. Australia **36**, e037 (2019)
13. Zwitter, T., Munari, U.: An Introduction to Analysis of Single Dispersion Spectra with IRAF, vol. 1. Osservatori Astronomici di Padova e Asiago, Dipartimento di Astronomia dell ..., Padova (2000)
14. Vaswani, A., Shazeer, N., Parmar, N., Uszkoreit, J., Jones, L., Gomez, A.N., Kaiser, Ł., Polosukhin, I.: Attention is all you need. Adv. Neural Inf. Process. Syst. **30** (2017)
15. Riggi, S., Magro, D., Sortino, R., De Marco, A., Bordiu, C., Cecconello, T., Hopkins, A.M., Marvil, J., Umana, G., Sciacca, E., Bufano, F., Ingallinera, A., Fiameni, G., Spampinato, C., Zarb Adami, K.: Astronomical source detection in radio continuum maps with deep neural networks. Astron. Comput. **42**, 100682 (2023)
16. Carion, N., Massa, F., Synnaeve, G., Usunier, N., Kirillov, A., Zagoruyko, S.: End-to-end object detection with transformers. In: European Conference on Computer Vision, pp. 213–229. Springer, Berlin (2020)
17. Dosovitskiy, A., Beyer, L., Kolesnikov, A., Weissenborn, D., Zhai, X., Unterthiner, T., Dehghani, M., Minderer, M., Heigold, G., Gelly, S., et al.: An image is worth 16x16 words: transformers for image recognition at scale (2020, preprint). arXiv:2010.11929
18. Lin, T.Y., Maire, M., Belongie, S., Hays, J., Perona, P., Ramanan, D., Dollár, P., Zitnick, C.L.: Microsoft Coco: common objects in context. In: European Conference on Computer Vision, pp. 740–755. Springer, Berlin (2014)

Radio Image Segmentation with Autoencoders

Hattie Stewart, Jason Yeung, and Mark Birkinshaw

Abstract We present an autoencoder neural network architecture as a method of source detection in radio continuum images. We use SKA Data Challenge datasets to test the model and recover 65% of the sources within the test dataset.

1 Introduction

The Square Kilometre Array (SKA) will be the world's largest radio interferometer. It promises extremely large volumes of data leading to new science opportunities. New methods of image processing and analysis will be required for handling the datasets. The SKA Observatory (SKAO) have released simulated model datasets for the radio astronomy community to analyse. These datasets are considered to be representative of future SKA data. In this work, we use the radio continuum images released in Science Data Challenge 1 (SDC1) [1]. The objective is to detect, classify, and measure the properties of the entire source population within the data. Since these datasets are large we apply Machine Learning (ML) techniques as fast and optimised methods of source extraction.

For this task we must identify areas of interest within each image and classify them, thus this can be considered an image segmentation task. We directly implement the Convolutional Autoencoder (CAE) [2] as a segmentation method for source extraction and illustrate some of the results.

H. Stewart (✉) · M. Birkinshaw
University of Bristol, HH Wills Physics Laboratory, Bristol, United Kingdom
e-mail: hattie.stewart@bristol.ac.uk

J. Yeung
SciML, Scientific Computing Department, Research Complex at Harwell, Rutherford Appleton Laboratory, Didcot, United Kingdom

F. Bufano et al. (eds.), *Machine Learning for Astrophysics*, Astrophysics and Space Science Proceedings 60, https://doi.org/10.1007/978-3-031-34167-0_28

2 Method

The SDC1 data contains 9 images, each at a different observation frequency (560 MHz, 1.4 GHz, 9.2 GHz) and of different duration (8, 100, 1000 hours). Corresponding truth catalogues are provided for each frequency, containing location, shape and size information. We create our training set by generating 60×60 pixel size cutouts from the 8 hour observation at 560 MHz and normalise between 0 and 1. The cutout size was selected for optimum performance of the CAE and is sufficiently large to account for the majority of sources in the data. We use the training catalogue provided to generate corresponding segmentation maps, filtering the catalogue sources by size. A segmentation map is a labelled image of the same dimensions as the cutout, each pixel having a numerical classification. We use binary segmentation for this case, using 0 to represent background and 1 to represent a source, placing all sources in a single class. The segmentation maps are generated by creating a 60×60 pixel size zero array. The zero array is updated to 1 where there are sources, which we locate from the truth catalogue. We refer to the binary segmentation map and the data cutout as data-map pairs, which we use to train the network.

We split our data-map pairs 80:10:10 for training, validation and test. During training the input data are reduced to a low dimensionality latent space, where source properties can be captured. The image is then reconstructed to the original input size. We train by minimising the loss between the input data and the output segmentation, thus the network learns to segment sources and background. We parse new cutouts to the trained network to predict their segmentation and use a Gaussian Mixture Model (GMM) for component analysis to recover a binary segmentation map comparable with the ground truth. See Fig. 1 for some examples.

3 Results

The network has proven accurate at detecting the location of sources with a high surface brightness, detecting 65% of the source population within the test set. It appears that the source population that remains undetected are those whose signal is close to the background. Further investigation is required to quantify to what signal/noise ratio we can detect sources. In Fig. 1 we can see that the network has struggled to capture the shapes of the sources accurately. In cutout 39 it appears that the network has predicted a source that does not occur in the ground truth; this source may have been cleaned during pre-processing and must be investigated further. Whilst we have yet to quantitatively compare the performance of the CAE with the source detection methods used for SDC1, namely PyBDSF [4] and ProFound [5], we have seen qualitatively that it performs at a similar level. We have since applied U-Net [3] to larger cutouts with the same method, trained on data-map pairs, which has yielded far better source recovery and shape reconstruction. Please see conference proceedings for IAU Symposium 368 for further information [6].

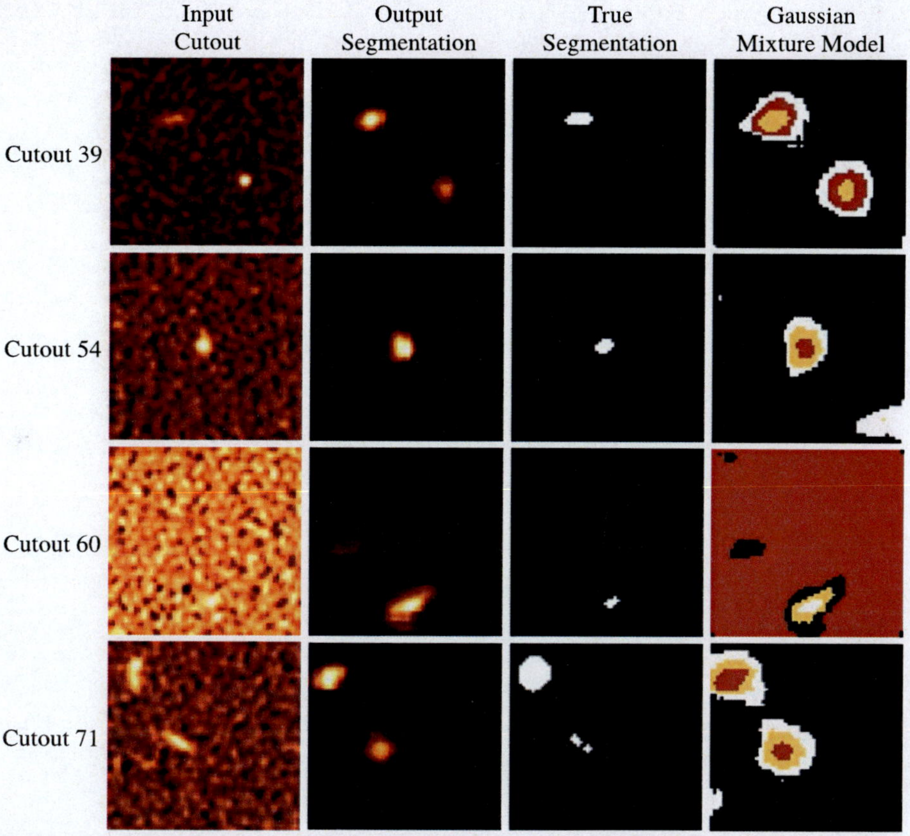

Fig. 1 Example cutouts from the test dataset demonstrate the performance of the network. 'Input Cutout' is the raw data cutout parsed to the network for prediction. 'Output Segmentation' is the segmentation map predicted by the trained network. 'True Segmentation' is the ground truth segmentation generated from the truth catalogue. 'Gaussian Mixture Model' is the output from a GMM applied to the output segmentation

References

1. Bonaldi, A., An, T., Brüggen, M., et al.: Square kilometre array science data challenge 1: analysis and results. Mon. Not. R. Astron. Soc. **500**, 3821–3837 (2021)
2. Dalca, A.V., Guttag, J., Sabuncu, M.R.: Anatomical priors in convolutional networks for unsupervised biomedical segmentation. In: Proceedings of the IEEE Computer Society Conference on Computer Vision and Pattern Recognition (2018)
3. Ronneberger, O., Fischer, P., & Brox, T.: U-Net: convolutional networks for biomedical image segmentation. In: Medical Image Computing and Computer-Assisted Intervention. LNCS. 9351, p. 234 (2018)
4. Mohan, N., Rafferty, D.: Python blob detection and source finder. Astrophysics Source Code Library, ascl:1502.007 (2015)

5. Robotham, A.S.G., Davies, L.J.M., Driver, S.P., et al.: ProFound: source extraction and application to modern survey data. Mon. Not. R. Astron. Soc. **476**, 3137–3159 (2018)
6. Stewart, H., Birkinshaw, M., Yeung, J.: Deep radio image segmentation. In: IAU Symposium Proceedings (in press) (2022)

Citizen Science and Machine Learning: Towards a Robust Large-Scale Automatic Classification in Astronomy

Manuel Jiménez, Emilio J. Alfaro, and Isaac Triguero

Abstract Citizen science is gaining momentum as a valuable resource for the classification of large collections of astronomical images by amateurs, although at the expense of worse quality in comparison with professional astronomers. To alleviate this issue, the inclusion of machine learning techniques in pursuit of a robust large-scale automated classification is considered mandatory. However, current approaches have disregarded the availability of expert knowledge along with vast amounts of unlabelled data that have not been leveraged yet within a unified learning framework. Here we present the Citizen Science Learning, an innovative algorithm capable of learning from unlabelled, amateur- and expert-labelled images employing a convolutional autoencoder and the pre-training and fine-tuning of a convolutional neural network, respectively. Our results demonstrate a better use of the label sets available in citizen science projects with respect to a set of comparative approaches in three classification scenarios from the *Galaxy Zoo* data.

1 Introduction

Citizen science (CzS) [1], traditionally known as the engagement of amateur participants in research, is demonstrating a great potential for large-scale processing of data. Using the power of the web, virtual communities of volunteers have been able to coordinate the classification of hundreds of thousands of images in a reasonable amount of time [2]. In areas such as astronomy or geo-sciences, where emerging technologies generate huge volumes of data, this approach entails image

M. Jiménez (✉) · E. J. Alfaro
Instituto de Astrofísica de Andalucía (CSIC), Granada, Spain
e-mail: mjimenez@iaa.es

I. Triguero
School of Computer Science, University of Nottingham, Nottingham, United Kingdom

Andalusian Research Institute in Data Science and Computational Intelligence (DaSCI), University of Granada, Granada, Spain

F. Bufano et al. (eds.), *Machine Learning for Astrophysics*, Astrophysics and Space Science Proceedings 60, https://doi.org/10.1007/978-3-031-34167-0_29

classification at a rate not possible to accomplish by experts alone, although at the expense of worse quality in the classifications made by amateur participants [3]. Despite its success in astronomy, as evidenced by the numerous editions of the *Galaxy Zoo* project [4], the current and upcoming massive surveys highlight its limitations, and the inclusion of machine learning methods towards a more robust automatic classification is considered mandatory [5]. However, current efforts attempting the exploitation of CzS outcomes with machine learning tools have ignored their inherent uncertainty as well as the potential of expert classifications to ameliorate this issue [6]. Their ultimate goal has mainly been to replicate the amateur performance, thus propagating their biases and limitations and disregarding the fact that, apart from the data labelled by amateurs, there is also available (limited) expert knowledge of the problem along with vast amounts of unlabelled data that have not been exploited yet within a unified learning framework.

Our research delves into the development of automated approaches for astronomical classification problems that have been aided by CzS projects on the web, aiming to leverage the inherent uncertainty in their results and all levels of knowledge available about the problem. We introduce an innovative learning paradigm for CzS projects in astronomy capable of taking advantage of expert- and amateur-labelled images, and unlabelled images. As an implementation of this learning framework, we present the Citizen Science Learning (CzSL), an algorithm that first learns from unlabelled data with a convolutional autoencoder (CAE), and then exploits amateur and expert labels via the pre-training and fine-tuning of a convolutional neural network (CNN), respectively, and a multi-layer perceptron (MLP). As a case study, we focus on the data derived from the first edition of the *Galaxy Zoo* project (GZ1) [7], from which we test binary, multi-class, and imbalanced classification scenarios. Our results demonstrate an improved classification performance in comparison to a representative set of baseline approaches, showing a more comprehensive use of these resources that are available to the astronomical research community.

The next Sect. 2 introduces the CzSL learning algorithm. In Sect. 3, we present the experiments conducted for its evaluation using the GZ1 data and results, and Sect. 4 summarises the final remarks.

2 Methodology

The CzSL algorithm learns in a bottom-up fashion, from the lowest level of knowledge to the highest, employing unlabelled data, amateur-labelled data, and expert-labelled data in three stages. In the following, we define train and test sets (TR and TS, respectively). Within the TR data, part of the data is considered unlabelled (U), and the rest is labelled by amateurs (A). For part of the A set, we also hold expert labels (E). That is to say: $U \cup A = TR$, and $E \in A$.

- **Stage 1**: CAE-CNN pre-training. CAE is trained using the whole TR data (no labels), and weights are transferred to the CNN. For this, the feature extraction layers of the CNN must resemble the CAE structure.
- **Stage 2**: CNN pre-training. The pre-trained CNN is pre-trained again using the A subset, i.e. the part of the TR data with amateur labels. The CNN weights are transferred to the next stage of the learning process.
- **Stage 3**: MLP extension—CNN fine-tuning. First, the expert knowledge in E is extended: a MLP is trained on E and predicts *enhanced* amateur labels for the rest of examples in A. Then, the CNN is fine-tuned using the enhanced amateur labels and expert labels in A set.

3 Experiments and Results

We carry out the experiments over three data sets from the original GZ1 results (Table 1): GZ1-SE, for binary classification (elliptical/spiral); GZ1-RLE, for multi-class classification (right-handed spiral/left-handed spiral/elliptical); and GZ1-MG, for imbalanced classification (merger/No-merger). In addition, we employ the GZ1-MG data set in conjunction with random over sampling (ROS) and random under sampling (RUS) techniques. The data sets are partitioned implementing a fivefold cross-validation, establishing 80% for TR set and 20% for TS set. Within TR, we consider 5% of TR as E, 20% as A, and the resulting 80% as U.

To evaluate the performance of our CzSL proposed approach, we define four comparative approaches. Amateur and Expert oracles train the CNN using the whole TR with amateur and expert labels, respectively. The MLP approach trains the CNN using expert labels in E and MLP predictions for the rest of examples in A. MLP-SL approach first performs the latter CNN training, predicts the labels for the rest of examples in U, and then completes a second training using expert labels in E, enhanced amateur labels in the rest of examples in A, and CNN predictions in U.

The results for the accuracy in TS are presented in Table 2. As it is shown, the CzSL is able to improve the performance in most of the problems, and MLP-SL also obtains even a better accuracy in GZ1-SE and GZ1-MG with ROS.

Table 1 Data sets used in the experiments

Data set	Problem	No. examples	Classes distribution (%)	
			Expert labels	Amateur labels
GZ1-SE	Binary	40,902	60.7:39.3	58.6:41.4
GZ1-RLE	Multi-class	40,902	30.1:30.6:39.3	28.8:29.9:41.3
GZ1-MG	Imbalanced	43,779	6.5:93.5	6.5:93.5

Table 2 Accuracy in TS obtained by comparative approaches and CzSL algorithm

Approach	GZ1-SE	GZ1-RLE	GZ1-MG	GZ1-MG + ROS	GZ1-MG + RUS
Amateur oracle	0.9169	0.8065	0.5768	0.7669	0.8754
Expert oracle	0.9510	0.8181	0.7592	0.8487	0.8738
MLP	0.9402	0.6585	0.6957	0.8333	0.8439
MLP-SL	0.9430	0.6656	0.7053	0.8373	0.8573
CzSL	0.9405	0.7192	0.7431	0.8144	0.8631

4 Conclusions

The CzSL approach has demonstrated to exploit the whole labelling framework deployed around the development of citizen science projects on the web involving astronomical image classification, showing a better performance in comparison with other comparative approaches. The pre-training and fine-tuning of deep learning-based classifiers offer a promising solution to learn from different confidence levels in the labels, which will be explored in future work using other data types such as tabulated data, spectra, etc.

Acknowledgments Authors Manuel Jiménez and Emilio J. Alfaro acknowledge financial support from the State Agency for Research of the Spanish MCIU through the "Center of Excellence Severo Ochoa" award to the Instituto de Astrofísica de Andalucía (SEV-2017-0709).

References

1. Show, H.: Rise of the citizen scientist. Nature **524**, 265 (2015). https://doi.org/10.1038/524265a
2. Masters, K.L.: Twelve years of Galaxy Zoo. Proc. Int. Astron. Union **14**, 205–212 (2019). https://doi.org/10.1017/S1743921319008615
3. Kosmala, M., Wiggings, A., Swanson, A., Simmons, B.: Assessing data quality in citizen science. Front. Ecol. Environ. **14**(10), 551–560 (2016). https://doi.org/10.1002/fee.1436
4. Lintott, C.J., Schawinski, K., Slosar, A., et al.: Galaxy Zoo: morphologies derived from visual inspection of galaxies from the Sloan Digital Sky Survey. Mon. R. Astron. Soc. **389**, 1179–1189 (2008). https://doi.org/10.1111/j.1365-2966.2008.13689.x
5. Walmsley, M., Lintott, C., Géron, T., et al.: Galaxy Zoo DECaLS: Detailed visual morphology measurements from volunteers and deep learning for 314 000 galaxies. Mon. Not. R. Astron. Soc. **509**, 3966–3988 (2022). https://doi.org/10.1093/mnras/stab2093
6. Dieleman, S., Willett, K.W., Dambre, J.: Rotation-invariant convolutional neural networks for galaxy morphology prediction. Mon. R. Astron. Soc. **450**(2), 1441–1459 (2015). https://doi.org/10.1093/mnras/stv632
7. Lintott, C., Schawinski, K., Bamford, S., et al.: Galaxy Zoo 1: Data release of morphological classifications for nearly 900 000 galaxies. Mon. R. Astron. Soc. **410**(1), 166–178 (2011). https://doi.org/10.1111/j.1365-2966.2010.17432.x

Background Estimation in Fermi Gamma-Ray Burst Monitor Lightcurves Through a Neural Network

Riccardo Crupi

Abstract The aim of this work is to propose a Machine Learning approach to estimate a background model for the Gamma-Ray Burst Monitor (GBM) of the Fermi satellite. It is employed a Neural Network (NN) to estimate each detector background signal given the information of the satellite and its detectors. The estimated background can be employed into a triggering algorithm to discover significant long/weak events that are not detected by other approaches. It is shown the potential of the model by estimating the background on GBM data for a Gamma-Ray Burst (GRB) present in the GBM catalog, the ultra-long GRB 091024. The proposed approach is straightforwardly generalizable to estimate the background model of other satellites.

1 Introduction

Gamma-Ray Bursts (GRBs) originate in extraordinarily energetic explosions taking place in distant galaxies. They are detected as irregular pulses of X and γ-ray radiation in high-energy detectors by satellites such as BATSE, SWIFT, Fermi Gamma-ray Space Telescope and the forthcoming HERMES-TP/SP [3]. The Fermi Gamma-ray Space Telescope hosts the Gamma-ray bursts Monitor (GBM) and Large Area Telescope (LAT) experiments. GBM contains 12 NaI detectors (efficient from 8 keV to 1 MeV) [2]. In this work Fermi GBM Data Tools [4] is employed.

To trigger an event an algorithm should decide if the foreground counts is statistically higher than the background counts. It is referred to *foreground* to indicate the counts of the photons received by the detector and to *background* to indicate the expected counts in the absence of γ/X-ray events. The *background* consists in different physical components [1, 6], that are not know a priori, and cannot be measured in isolation with respect to γ/X-ray events, thus an estimation

R. Crupi (✉)
University of Udine, Udine, Italy
e-mail: crupi.riccardo@spes.uniud.it

F. Bufano et al. (eds.), *Machine Learning for Astrophysics*, Astrophysics and Space Science Proceedings 60, https://doi.org/10.1007/978-3-031-34167-0_30

is needed to "clean" the *foreground* counts factoring out the *background*. The Fermi on-board trigger algorithm computes the *background* as an average counts of previous seconds [6] but it might fail to detect slowly rising events [1]. Employing a physical model to estimate the *background*, such as in [1], leads to good estimates and could reveal long/weak signals. In this study, it is demonstrated how, by using a Machine Learning approach, can be achieved equivalent findings with increased flexibility because, at least in principle, it can be applied to any satellite data.

2 Objective

The detectors considered are only the twelve NaI: *n0*, *n1*, *n2*, *n3*, *n4*, *n5*, *n6*, *n7*, *n8*, *n9*, *na*, *nb*. In this work, since the major concern are long GRBs, the counts lightcurves are binned at 4.096 s and binned in energy range of 3 categories: 28–50 keV (*r0*), 50–300 keV (*r1*), 300–500 keV (*r2*). In total there are 36 variables that are called *col_range*, 12 counts rate (in 4.096 s) times 3 energy range.

Thanks to the Fermi Data Tools package it is possible to retrieve information of the satellite in a particular timestamp. These variables are 24 (orbital information of Fermi, earth and sun location, etc.) and are referred to them as *col_sat_pos*. Information relative to the detector is also available. In total the variables are 36 (pointing of the detectors and FOV Earth occlusion) and are called *col_det_pos*.

It is formulated the problem as a typical supervised Machine Learning estimation, where input X are the variables in *col_sat_pos* and *col_det_pos* and the output Y the variables in *col_range*. It is estimated the photons in absence of events by means of the following formulation: the model requires to estimate an F such that $Y \approx F(X)$. Here the problem faced is a multi-output regression. The model employed is a feed forward neural network [5] with 3 hidden dense layers and 3 dropouts, one per each hidden layer.

The data in input where standardised, the instances in South Atlantic Anomaly were deleted. The splitting procedure divides the dataset in 75% train, 25% test; 30% of the training set is further kept as validation set. The loss function is the Mean Absolute Error $\mathrm{MAE}(x, y) = \frac{1}{n}\sum_{i=1}^{n}|y_i - x_i|$. In the settings of multi-output regression, the overall loss is $L = \min_i(\mathrm{MAE}(f_i(X), Y_i))$, where i spans *col_range*.

3 Results

To present a benchmark in terms of *background* estimation, it is analysed a time interval related to the ultra-long GRB 091024. Figure 1 shows the lightcurve for detector *n0* in range *r1*. Data are displayed in black, while red stands for the *background* estimation of a Neural Network trained and tested over a period of period of 3 months, from September to November 2009, that consists of 1.63 million

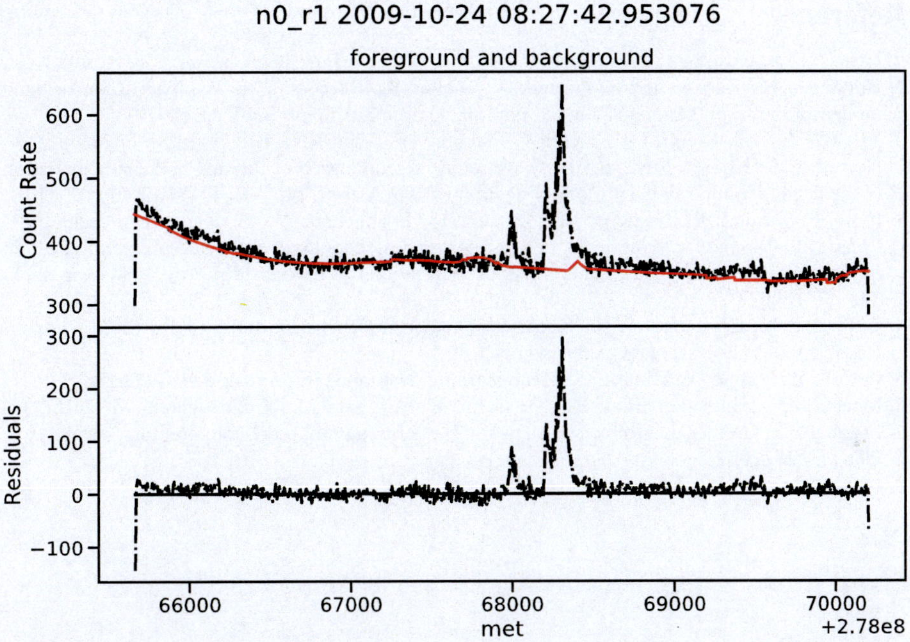

Fig. 1 *Foreground* and *background* fit around the event GRB 091024 for detector *n0* and range *r1*. This figure can be compared with figure 18 of [1]

of samples. The event emerges clearly from the residuals and the *background* fits well the shape of the lightcurve.

4 Conclusion

A Machine Learning estimation of the γ/X-ray *background* is introduced. Neural Network has been employed, leveraging on satellite data to perform a multi-output regression. The tests are performed on event GRB 091024. Results are in line with a physical *background* estimator, despite the approach is agnostic with respect to any physical assumption. This approach could be employed in recreating the *background* of other satellites, such as HERMES, then applying a trigger algorithm to efficiently recover astronomical transient events.

References

1. Biltzinger, B., Kunzweiler, F., Greiner, J., Toelge, K., Burgess, J.M.: A physical background model for the fermi gamma-ray burst monitor. Astron. Astrophys. **640**, A8 (2020)
2. Bissaldi, E., von Kienlin, A., Lichti, G., Steinle, H., Bhat, P.N., Briggs, M.S., Fishman, G.J., Hoover, A.S., Kippen, R.M., Krumrey, M., et al.: Groundbased calibration and characterization of the fermi gamma-ray burst monitor detectors. Exp. Astron. **24**(1–3), 47–88 (2009)
3. Fiore, F., Burderi, L., Lavagna, M., Bertacin, R., Evangelista, Y., Campana, R., Fuschino, F., Lunghi, P., Monge, A., Negri, B., et al.: The hermestechnologic and scientific pathfinder. In: Space Telescopes and Instrumentation 2020: Ultraviolet to Gamma Ray, vol. 11444, pp. 214–228. SPIE (2020)
4. Goldstein, A., Cleveland, W.H., Kocevski, D.: Fermi GBM data tools: v1.1.0 (2021). https://fermi.gsfc.nasa.gov/ssc/data/analysis/gbm
5. LeCun, Y., Bengio, Y., Hinton, G.: Deep learning. Nature **521**(7553), 436–444 (2015)
6. Meegan, C., Lichti, G., Bhat, P.N., Bissaldi, E., Briggs, M.S., Connaughton, V., Diehl, R., Fishman, G., Greiner, J., Hoover, A.S., et al.: The fermi gamma-ray burst monitor. Astrophys. J. **702**(1), 791 (2009)

Machine Learning Investigations for LSST: Strong Lens Mass Modeling and Photometric Redshift Estimation

Stefan Schuldt

Abstract Strong lensing analyses and photometric redshifts are both integral for cosmological studies with the Rubin Observatory Legacy Survey of Space and Time (LSST). With convolutional neural networks, we obtain significant gain both in performance and speed. We will present our new achievements in the HOLISMOKES collaboration in modeling galaxy-scale lenses using a residual neural network that was trained on very realistic mocks. This network enables us to predict the mass model parameters values with uncertainties, allowing us to predict in advance the next appearing image and corresponding time delays in case of e.g., a lensed supernova. We further present our dedicated model comparison of 31 real lenses, once obtained with the network and once with traditional techniques, for which we developed an automated code to drastically minimize the user input time. Furthermore, we will highlight our method NetZ to predict the photo-z based only on the pure galaxy images. Both networks are able to estimate the parameter values in fractions of a second while the lens modeling with non-automated traditional techniques takes typically weeks to months. With both networks and also our automated traditional pipeline, we are ready to process the huge amount of images obtained with LSST.

1 Modeling of Strongly Lensed Galaxies

Gravitational lensing means that the light of a background object such as a galaxy is gravitationally deflected by a massive object in the foreground, for instance another galaxy. Observations of this effect gives us the opportunity to probe the Universe in various aspects, including the study of high-redshift systems, the study of the nature and distribution of dark matter, and to carry out cosmological parameter measurements.

S. Schuldt (✉)
Max-Planck Institute for Astrophysics/Technical University of Munich, Munich, Germany
e-mail: schuldt@mpa-garching.mpg.de

F. Bufano et al. (eds.), *Machine Learning for Astrophysics*, Astrophysics and Space Science Proceedings 60, https://doi.org/10.1007/978-3-031-34167-0_31

Since the chance of a strong lensing event, which means that the background object is multiple observed, is relatively low, huge effort is spent on large lens detection programs. After a lens system is found, a mass model is necessary for nearly all further studies. Lens models are often described through parameterized profiles such as a Sérsic profile [1, 2] for the lens light. For the mass distribution, the typically adopted profiles depend on the image resolution and the scientific goal. For ground-based images of galaxy-scale lensing systems as considered here, one typically adopts a singular isothermal ellipsoid (SIE) profile, possibly together with an external shear component with strength $\gamma_{\rm ext}$.

1.1 Automated Traditional Modeling: glee_auto.py and glee_tools.py

The parameter values that describes best the observations are often obtained through iterative runs of Markov-Chain Monte Carlo chains, which is a very time and resource consuming procedure. This means, modeling of a single lens can take up to several weeks. Given the increasing number of lens detection thanks to current and upcoming wide field imaging surveys like the Rubin Observatory Legacy Survey of Space and Time (LSST), this procedure is not enough anymore.

One possibility is to automate the modeling procedure while still relying on Bayesian inference as this reduces the user input time already drastically. To this end, we developed *glee_auto.py* [3], an automated model procedure dedicated for galaxy-scale lensing systems perfectly suited for LSST-like ground based images. This pipeline consists out of four steps: (1) preparation of required input files, (2) lens light modeling, (3) a quick source and image position optimization (4) an arc light modeling together with the source surface brightness reconstruction. Further details can be found in [3].

Since each lensing system with its environment is unique, *glee_auto.py* does not work perfectly on all systems. Therefore, we presented in [3] also *glee_tools.py*, a flexible modeling code that already can run multiple optimization steps without further user input and is thus perfectly suited for follow-up model refinements independent of the lensing system or adopted profiles.

1.2 Modeling with Neural Networks

Nonetheless, the procedure remains with around days computationally very lengthy when taking into account the number of lens detections with ~100,000 from LSST. Another option is to use convolutional neural networks (CNNs) as proposed by [4]. The main advantage is the fully autonomous procedure and the extreme speed-up as the trained network predicts the mass parameters within fractions

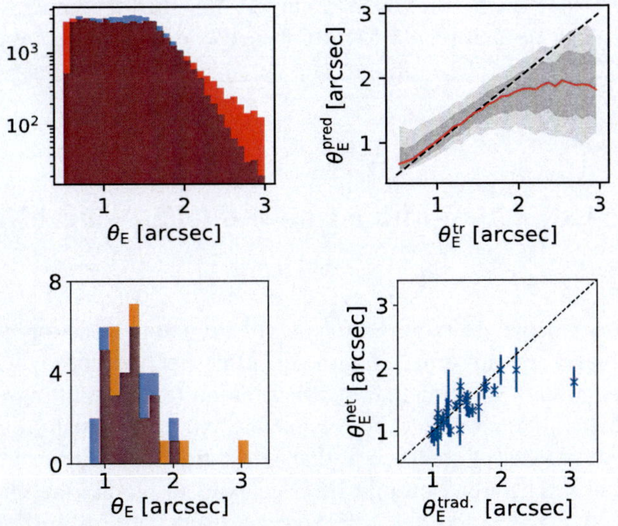

Fig. 1 Network performance on the Einstein radius in comparison to the test set (top) and to the traditionally obtained parameter values of real lenses (bottom). In both rows, we show on the left a histogram and on the right side the direct comparison with the network predictions on the y-axis

of a second on a single CPU. Therefore, we first presented a simple CNN to predict the SIE parameters in [5], which we then upgrade to predict the SIE+γ_{ext} parameters together with 1σ uncertainties [6] using a so-called residual neural network (ResNet) which is a specific type of CNN. Both networks are trained on realistic mock images that were generated with our simulation pipeline [5, 6] using images of real galaxies from the Hyper Suprime-Cam (HSC) matching good the quality expected from LSST. This is the main difference to other modeling networks, which always assume HST-like high-resolution images from space. The network performs good on the SIE parameters (see Fig. 1 for the Einstein radius), while difficulties remain on the external shear raised most likely from the low resolution. For further details see [5, 6].

1.3 Direct Comparison Between Neural Network and Traditional Modeling on Real Lenses

Since the network was trained and tested on mock images, the performance on real images is still unclear. To this end, we compared models of 31 real HSC lenses obtained (1) with the ResNet and (2) with *glee_auto.py* and *glee_tools.py*, i.e. the traditional method [3]. We show the comparison for the Einstein radius in Fig. 1 and find comparable performance for the other parameters as obtained with the test set.

This shows the power of CNNs requited for the upcoming data-rich era. We further present an automated MCMC-based code for more detailed analysis, dedicated for LSST-like observations. The details of the comparison can be found in [3].

2 Photo-z Estimation with a Convolutional Neural Network: NetZ

Redshift measurements are required for nearly any study of astrophysical objects. Since most objects are only imaged, large effort is spent on improving photometric redshift, either trough Spectral Energy Distribution template fitting or, nowadays, using neural networks. However, both types are typically based on characteristics such as the color, size, or magnitude that got extracted from the images. In [7], we presented a CNN trained on real HSC galaxies to predict the photo-z directly from the images. A detailed and direct comparison with DEmP [8], one of the best methods applied to HSC data, demonstrates that NetZ is competitive to other methods. While our main network $\text{NetZ}_{\text{main}}$ predicts redshifts in the range 0 to $\sim$4, we trained also $\text{NetZ}_{\text{lowz}}$ in the range 0 and 1 as well as NetZ_{LRG} specifically on luminous red galaxies, giving both good improvement for their dedicated sample.

We demonstrate with NetZ the possibility of photo-z predictions up to $z \sim 4$ using directly the HSC image stamps as preparation for LSST. For further details see [7].

Acknowledgments I thank R. Cañameras, S. Ertl, A. Halkola, B. C. Hsieh, L. Leal-Taixé, T. Meinhardt, Y. Shu, S. H. Suyu, and S. Taubenberger for the corresponding collaboration, the two anonymous referees, and C. Grillo for support. SS thanks the Max Planck Society for support through the Max Planck Research Group for S. H. Suyu. This project has received funding from the European Research Council (ERC) under the European Unions Horizon 2020 research and innovation programme (LENSNOVA: grant agreement No 771776). This research is supported in part by the Excellence Cluster ORIGINS which is funded by the Deutsche Forschungsgemeinschaft (DFG, German Research Foundation) under Germany's Excellence Strategy—EXC-2094—390783311.

References

1. De Vaucouleurs, G.: Sur la Loi de Distribution de la Luminosite dans les Nebuleuses Elliptiques et leur Structure. C. R. Hebd. Seances Acad. Sci. **227**, 586 (1948)
2. Sérsic, J.L.: Influence of the atmospheric and instrumental dispersion on the brightness distribution in a galaxy. Boletin de la Asociacion Argentina de Astronomia La Plata 2.Argentina, 6, 41, Feb. (1963)
3. Schuldt, S., Suyu, Cnameras, S.H., Shu, Y., Taubenberger, S., Ertl, S., Halkola, A.: HOLISMOKES - X. Comparison between neural network and semi-automated traditional modeling of strong lenses. Astron. Astrophys. **673**, A33 (2023). https://doi.org/10.1051/0004-6361/202244534

4. Hezaveh, Y.D., Perreault Levasseur, L., Marshall, P.J.: Fast automated analysis of strong gravitational lenses with convolutional neural networks. Nature **548**, 555 (2017)
5. Schuldt, S., Suyu, S.H., Meinhardt, T., Leal-Taixé, L., Cañameras, R., Taubenberger, S., Halkola, A.: HOLISMOKES IV. Efficient mass modeling of strong lenses through deep learning. Astron. Astrophys. **646**, A126 (2021)
6. Schuldt, S., Cañameras, R., Shu, Y., Suyu, S.H., Taubenberger, S., Meinhardt, T., Leal-Taixé, L.: HOLISMOKES - IX. Neural network inference of strong-lens parameters and uncertainties from ground-based images. Astron. Astrophys. **671**, A147 (2023)
7. Schuldt, S., Suyu, S.H., Cañameras, R., Taubenberger, S., Meinhardt, T., Leal-Taixé, L., Hsieh, B.C.: Photometric redshift estimation with a convolutional neural network: NetZ. Astron. Astrophys. **651**, A55 (2021)
8. Hsieh, B.C., Yee, H.K.C.: Estimating luminosities and stellar masses of galaxies photometrically without determining redshifts. Astrophys. J. **792**, 102 (2014)

Multi-Band Photometry and Photometric Redshifts from Astronomical Images

Laura Cabayol-Garcia, Martin Børstad Eriksen, and the PAUS Collaboration

Abstract The next generation of galaxy surveys, e.g. *Euclid* and Large-aperture Synoptic Survey Telescope (LSST), will drastically increase the volume and quality of astronomical data. To process and analyse the data will require fast and precise methods, including accurate measurements of the flux and redshift of galaxies. `Aczio` is an end-to-end deep-learning pipeline to simultaneously measure the co-added multi-band photometry and the photometric redshift of a galaxy directly from multiple astronomical images. `Aczio` uses all the image observations of a galaxy to measure the photometry in each photometric band, which enables to use the underlying galaxy spectral energy distribution (SED) to improve the galaxy flux measurements. `Aczio` doubles the signal-to-noise of the photometry compared to aperture-photometry measurements and predicts significantly better photo-zs than existing methods for $z > 1$.

1 Introduction

Imaging galaxy surveys require precise photometry and photometric-redshift measurements (photo-zs) for an extensive set of science applications, e.g. weak gravitational lensing, galaxy clustering, and intrinsic alignments. In recent years, the amount and quality of galaxy-survey data have drastically increased, requiring more precise and efficient methods for data analysis. `Aczio` is a multi-task deep neural network predicting the multi-band photometry and the photometric redshift of galaxies directly from the astronomical images. This enables a fast and precise evaluation of the photo-z, drastically reducing the data-reduction process while also having photometry measurements [1]. `Aczio` builds on the previous work in

L. Cabayol-Garcia (✉)
Institut de Física d'Altes Energies. Campus UAB, Facultat Ciencies Nord, Barcelona, Spain
e-mail: lcabayol@pic.es

M. B. Eriksen
Port d'Infoirmació Científica, Barcelona, Spain

F. Bufano et al. (eds.), *Machine Learning for Astrophysics*, Astrophysics and Space Science Proceedings 60, https://doi.org/10.1007/978-3-031-34167-0_32

Deepz [2] and Lumos [3], two deep neural networks predicting the photo-z and the single-epoch single-band photometry of galaxies, respectively.

2 Data

In this work, we use data from the Physics of the Accelerating Universe Survey (PAUS). PAUS is an imaging survey observing with a 40 narrow-band photometric filters camera (PAUCam, [4]) covering the optical wavelength range from 4500 to 8500 Å. The narrow-band filter set effectively measures a low-resolution spectrum ($R \approx 50$) enabling a photo-z precision of $\sigma_z = 0.0037(1+z)$ for a selected best 50% of the sample with $i_{AB} < 22.5$ with a template-fitting method [5]. This precision was improved by 50% with the deep neural network Deepz [2]. This study uses PAUS data in the COSMOS field [6]. These are the data used in the Lumos and Deepz papers and also in all the photo-z studies published so far, e.g. [2, 5, 7]. The complete photometry catalogue contains 64,476 galaxies to $i_{AB} < 23$ with 40 narrow-band filters. With approximately five observations per galaxy and narrow-band filter, this corresponds to around 12.5 million galaxy observations.

2.1 *Training Sample*

Since there are no PAUS data with known narrow-band photometry, we use image simulations to train the neural network. The Teahupoo simulations [3] are a set of single-observation PAUS-like galaxy images combining a randomly selected 60×60 PAUCam cutout and a bulge+disk simulated galaxy. To train the photo-z network, we also use the spectroscopic redshift sample compiled in [7]. This includes the zCOSMOS DR3 bright spectroscopic data [6] and a compilation of redshifts from several spectroscopic surveys.

3 Method

Aczio has two main parts: one extracting features from the galaxy images and another estimating the galaxy photometry and the photometric redshift from such features. To extract the features, we use a 2D convolutional neural network (CNN), which extracts 20 features per image. These 20 features encode the relevant image properties required for predicting the photometry and photo-z, e.g. the size, shape, ellipticity, total flux of the galaxy, background-noise properties, and the presence nearby artifacts. In order to combine the features of independent exposures of the same galaxy with the same photometric filter, we use an adaptive-pooling layer [8], which enables fixing the output size of the layer given any input and provides a flexible way of dealing with different number of observations per training galaxy. Since the network needs information from all the bands to estimate the photometric redshift of galaxies, we combine the features from all bands in a 800 features tensor. `Aczio` predicts the probability distribution of the photo-z and the photometry predictions from the 800 galaxy features using Gaussian Mixture Density Networks (MDNs). Ideally, these a single MDN would make all these predictions while accounting for correlations among bands and using all the photometry information to improve the photo-z. However, this is work in progress and so far we are using 40 independent MDNs to predict the photometry in each of PAUS narrow bands.

4 Results

The left panel in Fig. 1 shows the signal-to-noise of the co-added photometry for ≈14,000 PAUCam galaxies in ten equally-populated magnitude bins with the aperture-photometry method `MEMBA`, the `Lumos` deep neural network, and `Aczio` (this study). While `Lumos` independently measures the photometry in each band from a single-exposure image, `Aczio` can use the information encoded in all the images of a galaxy, which enables to use the underlying galaxy SED to improve the photometry measurements in the individual narrow bands. `Aczio` provides higher signal-to-noise than `MEMBA` and `Lumos` at all magnitudes. This is very significant at the faint end, where the signal-to-noise is three times higher than with aperture photometry and two times higher than with `Lumos`.

To evaluate the accuracy and precision of the photo-z estimates, we define

$$\Delta z := (z - z_{\mathrm{s}}) \;/\; (1 + z_{\mathrm{s}}) \,, \tag{1}$$

where z and z_{s} are the photo-z and the spectroscopic redshifts, respectively. The bias and the dispersion are defined as the median and σ_{68} of Δz, respectively, where we

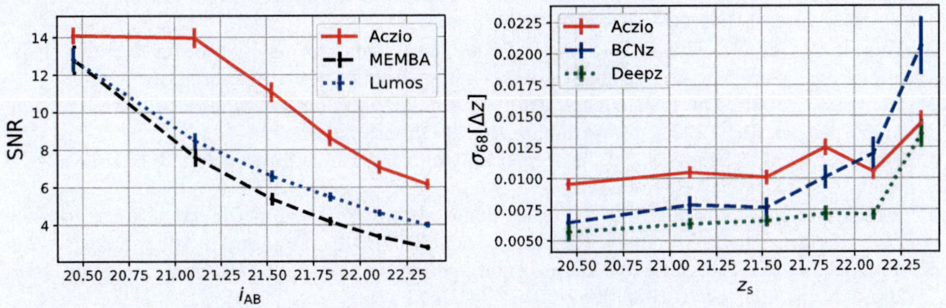

Fig. 1 *Left*: Signal-to-noise (SNR) of the photometry of ≈14,000 PAUCam galaxies with: `Aczio` (red solid), `Lumos` (dotted blue), and `MEMBA` (dashed black) in equally populated magnitude bins. *Right*: Photo-z precision in equally populated spectroscopic-redshift bins with `Aczio` (solid red), `BCNz` (dotted blue), and `Deepz` (dashed green)

define σ_{68} as

$$\sigma_{68} \coloneqq \frac{1}{2}\left[Q_{84}(\Delta z) - Q_{16}(\Delta z)\right] , \tag{2}$$

and $Q_{16}(\Delta z)$, $Q_{84}(\Delta z)$ are the 16th and 84th percentiles of the Δz distribution.

The right panel of Fig. 1 shows the photo-z precision in spectroscopic-redshift bins for `Aczio` (solid red), the template-fitting code `BCNz` (dotted blue), and the deep neural network `Deepz` (dashed green). For high-redshift galaxies, `Aczio` already provides better photo-z than the other two methods. However, the plot also reveals a problem at low redshift, where it does not reach the photo-z precision of `BCNz` and `Deepz`. Fixing this problem is still work in progress.

5 Conclusions

Imaging surveys require accurate and precise galaxy photometry and photometric redshifts for most scientific analysis. In this work, we present `Aczio`, a CNN to predict both the multi-band photometry and the galaxy redshift directly from multi-band image observations. To the best of our knowledge, simultaneously predicting both photometry and photo-z is a novel contribution of this work. `Aczio` improves the photo-z measurements of `BCNz` and `Deepz` for galaxies with $z > 1$, However, for lower-redshift galaxies `Aczio` is still not providing the expected photo-z precision. This is still work in progress and in our near-future actions, we aim to study the photo-z dependence with redshift to improve the photo-z precision.

References

1. D'Isanto, A., Polsterer, K.L.: Photometric redshift estimation via deep learning. Generalized and pre-classification-less, image based, fully probabilistic redshifts. Astron. Astrophys. (2018). https://doi.org/10.1051/0004-6361/201731326
2. Eriksen, M., et al.: The PAU Survey: Photometric redshifts using transfer learning from simulations. Mon. Not. R. Astron. Soc. (2018). https://doi.org/10.1093/mnras/staa2265
3. Cabayol, L., et al.: The PAU survey: estimating galaxy photometry with deep learning. Mon. Not. R. Astron. Soc. (2021). https://doi.org/10.1093/mnras/stab1909
4. Padilla, C., et al.: The physics of the accelerating universe camera. Astrophys. J. (2019). https://doi.org/10.3847/1538-3881/ab0412
5. Eriksen, M., et al.: The PAU Survey: early demonstration of photometric redshift performance in the COSMOS field. Mon. Not. R. Astron. Soc. (2019). https://doi.org/10.1093/mnras/stz204
6. Lilly, S.J., et al.: The zCOSMOS 10k-bright spectroscopic sample. Astrophys. J. Suppl. Ser. (2009). https://doi.org/10.1088/0067-0049/184/2/218
7. Alarcon, A., et al.: The PAU Survey: an improved photo-z sample in the COSMOS field. Mon. Not. R. Astron. Soc. (2021). https://doi.org/10.1093/mnras/staa3659
8. Liu, Y., Zhang, Y.M., Zhang, X.Y., Liu, C.L.: Adaptive spatial pooling for image classification. Pattern Recogn. (2016). https://doi.org/10.1016/j.patcog.2016.01.030

Inference of Galaxy Clusters' Mass Radial Profiles from Compton-y Maps with Deep Learning Techniques

A. Sbriglio, W. Cui, D. de Andres, M. De Petris, A. Ferragamo, and G. Yepes

Abstract Clusters of galaxies are crucial tracers of the history of structure formation, and their mass function at different epochs is of fundamental importance for constraining cosmological parameters. Therefore, it is essential to infer the mass of observed clusters, which unfortunately is not directly observable and is affected by several biases related to applied observational estimates. To overcome these obstacles, we developed a machine learning algorithm capable of inferring the 3D cumulative radial profiles of the total mass and gas mass in galaxy clusters from the Sunyaev-Zel'dovich effect heat maps. In recent work, convolutional neural networks (CNNs) were applied to full-sky Compton-y maps to estimate cluster masses defined at a fixed aperture radius corresponding to 500 times the critical overdensity. We now extend this study to estimate the radial profiles of the total mass and gas mass of the clusters. We produced about 73,000 images along various lines of sight using 2522 clusters simulated by project at redshift $z < 0.12$ and trained a model combining an autoencoder and a random forest. We selected a samples with masses in the range $10^{13.5} \leq M_{200}(h^{-1}M\odot) \leq 10^{15.5}$ and different dynamical states. The obtained model was able to reconstruct unbiased profiles with a dispersion of about 10%, slightly incremental toward the core and periphery of the cluster. We show that both the accuracy and precision of this method show a slight dependence on the dynamical state, but not on the mass of the cluster.

A. Sbriglio (✉) · A. Ferragamo
Dipartimento di Fisica, Sapienza Universita di Roma, Roma, Italy

W. Cui
Institute for Astronomy, University of Edinburgh, Edinburgh, United Kingdom

D. de Andres · G. Yepes
Departamento de Fisica Teorica, Modulo 8, Facultad de Ciencias, Universidad Autonoma de Madrid, Madrid, Spain

M. De Petris
Dipartimento di Fisica, Sapienza Universita di Roma, Roma, Italy

F. Bufano et al. (eds.), *Machine Learning for Astrophysics*, Astrophysics and Space Science Proceedings 60, https://doi.org/10.1007/978-3-031-34167-0_33

1 Scientific Context

Clusters of galaxies are the largest gravitationally bound structures in the Universe. They are mainly composed of Dark Matter, as well as baryons in the phase of hot ionized gas (Intra Cluster Medium) and finally stars in galaxies. Their physics is of primary importance in Cosmology, in particular, their abundance in the Universe and their evolution are sensitive to the cosmic matter density, Ω_m, and the current amplitude of density fluctuations characterized by the σ_8 parameter, the linear effective overdensity in spheres of radius $8h^{-1}Mpc$. It is mandatory to have a reliable estimate of the total mass of the cluster which, observationally, is inferred by several methods imposing some approximations. Since it can also be affected by systematics, the use of synthetic clusters, generated by hydrodynamic simulations in which every feature of the cluster is well known, is employed. Moreover observational methods, such as the inference under the assumption of the hydrostatic equilibrium (HE), result in a biased mass estimate. This bias, quantified in the case of simulated clusters, is of the order of 10–20% [1]. In this work, see also [2], we will show how the application of a modern analysis method based on the implementation of Machine Learning algorithms trained to infer mass profiles of clusters from Compton-y maps related to the Sunyaev Zel'dovich (SZ) [3] effect, has allowed us to estimate mass profiles of clusters with negligible bias.

2 Dataset Selection and Random Forest Training

The dataset used in this work is composed of 2522 radial profiles of mock galaxy clusters, extracted with the Amiga's Halo Finder [4] in simulation and about 73,000 y-maps, corresponding to 29 projections for each cluster along different l.o.s (line of sight), created through the public package PYMSZ [5, 6]. It was used to train a Random Forest, a supervised ML algorithm capable of performing regressions between a set features and a set of targets.

The selected clusters have a mass range $10^{13.5} \leq M_{200}/(h^{-1}M\odot) \leq 10^{15.5}$ and a redshift range $z \in [0, 0.116]$. They span different dynamical states, from very relaxed to very disturbed halos. We also included the radial profiles of the total gas mass of each cluster up to $r = 2R_{200}$, summing the mass of all the particles within concentric spheres, centred in the highest density peak. In order to homogeneously sample the profiles, we interpolated the mass profiles at 24 linearly spaced fixed overdensities Δ, i.e. the ratio between the cluster density within a sphere with radius R_Δ and the critical density of the Universe at the cluster redshift, from $\Delta = 200$ to $\Delta = 2500$. We extracted the latent vectors containing the peculiar features of the synthetic SZ maps of each cluster by means of Autoencoder, an unsupervised neural network able to derive a representative feature vector with a reduced dimension of our input data while being faithful to the original input. The size of the maps was set to 128×128 pixels to reduce the size and training computation time. Each map has a size of $2R_{200} \times 2R_{200}$ and the angular resolution is adapted consistently. The 29 different maps allow us to increase the statistics and infers 3D profiles.

The synthesized dataset was split into a training part (80%) and a testing part (20%). Each part contained a *features* portion (latent vectors) and a *targets* portion (interpolated profiles). Individual profiles were considered for every 29 projections and the algorithm was trained to infer different quantities: a single interpolated point per cluster, 6 points at a time corresponding to an entire profile or the set of 48 points, namely all types of profiles considered for each cluster. A calibration of the hyperparameters defining the qualities of the forest response according to the type of data was carried out, and the profiles returned in the various cases were analyzed. The analysis showed that the best compromise was to use about 200 trees as deep as possible (many decision nodes able of capturing all the intrinsic characteristics of the data). In addition, the best inference was obtained for entire profiles instead of individual points, as the algorithm is thus able to capture correlations between various quantities.

2.1 Results

After planting and training the forest, the algorithm was applied on the feature component in the testing portion. The inferred profiles were so compared with the component of targets contained in the test by calculation of *bias*, defined as $b = \frac{M_{pred}}{M_{true}} - 1$. The total mass and gas mass profiles were then used to calculate the gas fraction f_g. In Fig. 1, we can see in blue the median bias for the total mass profiles and the respective scatters corresponding to the 16th–84th percentiles, in orange the same for the gas mass, and in green the case of the gas fraction.

What we get is a promising result that returns estimates with very low bias ($<1\%$) differently from mass inference based on HE approximation, and with errors on the order of 10% (M_{tot} and M_{gas}), which are reduced to as low as 3% (f_g). A detailed study of the dynamical state of the clusters showed how these scatters are affected by the relaxation state of the system, quantified as in [7]. Moreover, the gas fraction at $\Delta = 200$, $f_{g,200} = 0.18 \pm 0.01$, is compatible with Planck [8].

The algorithm developed in this work could also be applied to infer other radial profiles of the ICM, such as gas temperature and pressure. The reconstruction of the profiles, moreover, could hopefully be refined by employing the algorithm to different observational maps, such as optical and X-ray images. The next planned step is training on SZ mock maps, complete with instrumental and observational effects. Finally, a test on real maps of Compton-y parameters at different angular resolutions and redshifts is contemplated. The bias in the inferred ML mass will also be compared with other methods using simulated weak lensing images of the same dataset.

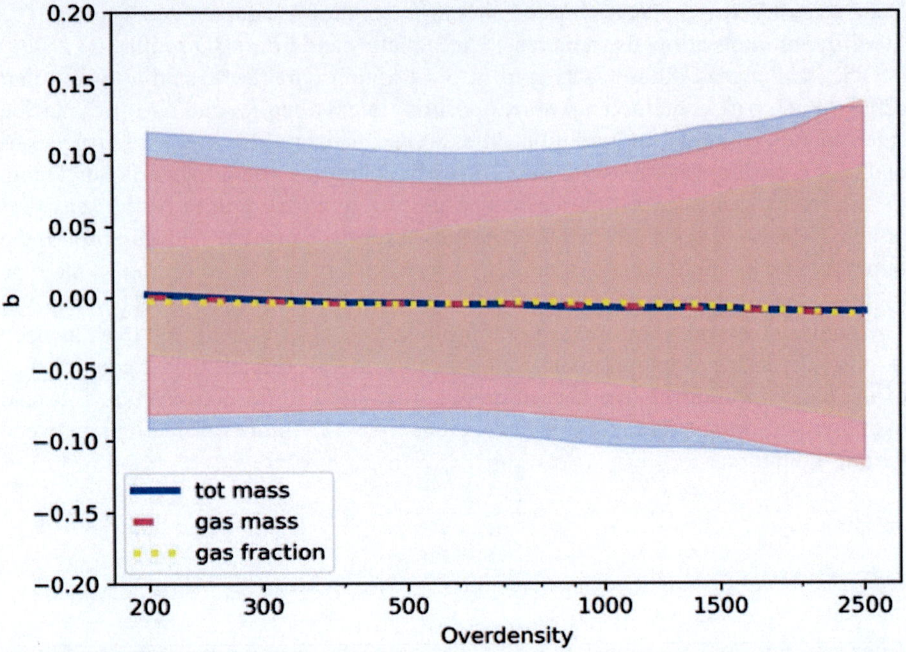

Fig. 1 The median bias as a function of the overdensity for total mass (blue solid line), gas mass (violet dashed line) and gas fraction (yellow dotted line). The shaded light blue, violet and yellow regions correspond to the 16th–84th percentiles for total, gas mass and gas fraction, respectively

References

1. Gianfagna, G., De Petris, M., Yepes, G., et al.: Exploring the hydrostatic mass bias in MUSIC clusters: application to the NIKA2 mock sample. Mon. Not. R. Astron. Soc. **502**, 5115 (2021)
2. Ferragamo, A., de Andres, D., Sbriglio, A., et al.: The Three Hundred project: a machine learning method to infer clusters of galaxies mass radial profiles from mock Sunyaev-Zel'dovich maps. Mon. Not. R. Astron. Soc. **22**, 2862 (2022). Published on 02 February 2023
3. Sunyaev R.A., Zeldovich Y.B.: Small-scale fluctuations of relic radiation. Astrophys. Space Sci. **7**, 3 (1970)
4. Knollmann, S.R., et al.: AHF: Amiga's halo finder. Astrophys. J. Suppl. Ser. **182**, 608–624 (2009)
5. Cui, W., Knebe, A., Yepes, G., et al.: The Three Hundred project: a large catalogue of theoretically modelled galaxy clusters for cosmological and astrophysical applications. Mon. Not. R. Astron. Soc. **480**, 2898 (2018)
6. Baldi, A., De Petris, M., Sembolini, F., et al.: Kinetic Sunyaev–Zel'dovich effect in rotating galaxy clusters from MUSIC simulations. Mon. Not. R. Astron. Soc. **479**, 4028 (2018)
7. De Luca, F., De Petris, M., Yepes, G., et al.: The Three Hundred project: dynamical state of galaxy clusters and morphology from multiwavelength synthetic maps. Mon. Not. R. Astron. Soc. **504**, 5383 (2021)
8. Planck Collaboration, Aghanim, N., et al.: Planck 2018 results-VI. Cosmological parameters. Astron. Astrophys. **641**, A6 (2020)

Deep Learning 21 cm Lightcones in 3D

Caroline Heneka

Abstract Interferometric measurements of the 21 cm signal are a prime example of the data-driven era in astrophysics we are entering with current and upcoming experiments. We showcase the use of deep networks that are tailored for the structure of 3D tomographic 21 cm light-cones to firstly detect and characterise HI sources and to secondly directly infer global astrophysical and cosmological model parameters. We compare different architectures and highlight how 3D CNN architectures that mirror the data structure are the best-performing models.

1 21 cm Imaging in 3D to Track the History of the Universe

The 21 cm signal originates from the forbidden spin-flip transition of neutral hydrogen. As hydrogen is the most abundant element in the Universe it is thus an optimal tracer of structure formation across cosmic time. At high redshift we can map heating due to the very first galaxies during cosmic dawn (CD), followed by the Epoch of Reionisation (EoR) when neutral hydrogen is ionised again. But also late-time structure formation and galaxy evolution is traced by the 21 cm signal via neutral hydrogen within sources. Amazingly, the Square Kilometre Array (SKA, https://www.skao.int/) will achieve 3D tomography and thus produce as data lightcones of 2D spatial maps and finely gridded frequency information. SKA-MID at frequencies of 350 MHz to 15 GHz detects HI sources post-reionisation, SKA-LOW targets at lower frequencies the high-redshift Universe. Due to up to TB/s of data inflowing from SKA, and a highly non-Gaussian signal, deep learning

C. Heneka (✉)
Institute for Theoretical Physics, Heidelberg University, Heidelberg, Germany
e-mail: heneka@thphys.uni-heidelberg.de

F. Bufano et al. (eds.), *Machine Learning for Astrophysics*, Astrophysics and Space Science Proceedings 60, https://doi.org/10.1007/978-3-031-34167-0_34

is suitable for the challenge to reduce and analyse this data. Important goals are both source detection for clustering studies and inference of global astrophysical and cosmological parameters. We will highlight here our findings regarding 21 cm tomography and optimal deep learning models for both these tasks.

2 Low-Redshift in 3D: 21 cm Source Detection

A 3D data cube as expected for the SKA was presented as a 1TB cube for the second SKA science data challenge (SDC2, https://sdc2.astronomers.skatelescope.org) for detection and characterisation of >100,000 HI sources (galaxies) centred at a redshift of $z = 0.235 - 0.495$ [1]. Together with the full cube a 1GB cube with ground truth information was provided for training.

Our team 'coin' optimised a couple of modern machine learning algorithms for detection (segmentation) of sources in the SDC2 HI cube and characterisation of source properties. These were integrated HI flux, line width in frequency, HI spatial size and inclination as well as position angle. In addition, a 'classical' detection algorithm based on wavelet filtering for de-noising and segmentation was compared to as a baseline. We considered the following architectures for object detection: 2D and 3D U-Nets, [2] variants of R-CNN (region-proposal convolutional neural network [3]) and an inception-style network [4] that mimics filtering with wavelets. Main challenges we identified, where data volume is not a problem for a trained network, are (1) low signal-to-noise, (2) small spatial size (few pix), (3) unknown systematics (f. ex. residual beam effects, radio frequency interference, RFIs).

We found that a shallow segmentation U-Net architecture generalised to 3D performed best. Trained on 3D cubic patches that each contain a source this architecture is able to provide a 3D reconstruction of HI sources without any pre-processing. Subsequent shallow 3D CNN regression models applied to the high-quality reconstructed images allowed for low-bias and high precision recovery of flux and HI size, almost independently of source properties such as brightness. Without pre-processing rates of false positives reached moderate levels ($\sim$50%), which can be further mitigated by noise modelling and pre-processing, training on an improved (higher signal-to-noise) fiducial training set, and an extra network-based decision step. Figure 1 showcases such a reconstruction; key points to emphasise are:

- 3D networks perform best for 3D segmentation of HI sources,
- 3D U-Nets provide high-fidelity 3D reconstructions,
- which allows for recovery of properties such as HI flux and size *almost independent of source brightness*.

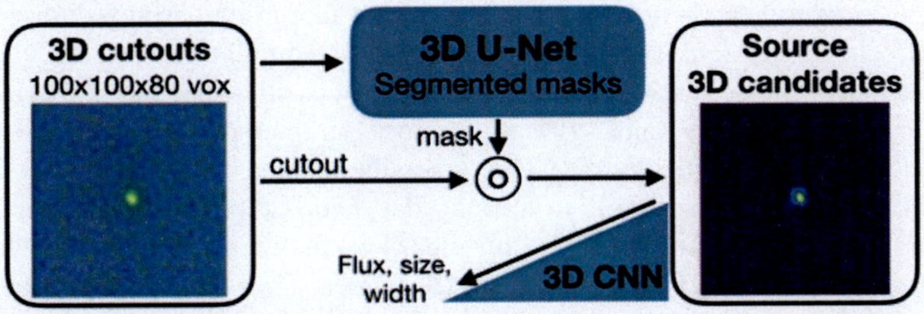

Fig. 1 Schematic of our 3D source detection and reconstruction, and HI source property derivation

3 High-Redshift in 3D: 21 cm Lightcone Inference

Moving from low-redshift to high-redshift measurements of the 21 cm signal, we showcase direct likelihood-free inference from 3D tomography during CD and EoR. To infer a set of astrophysical and cosmological parameters,[1] we have different options to treat 3D data optimally with network architectures:

- slicing for set of spatial images → 2D CNN, or e.g. ResNet with skip connections,[2]
- a time-series (frequency-series) of co-eval (spatial) images → LSTM network,[3]
- full 3D convolution → 3D CNN, moving from 2D to full 3D convolution.

We again found a shallow 3D CNN architecture as the best-performing model, in terms of low scatter and bias on inferred parameters, as well as fast and good convergence during training (low loss while avoiding overfitting). The 3D-21cmPIE-Net [4] consists of only four 3D-CNN layers, followed by a GAP (global average pooling) layer against overfitting and three densely connected layers to regress parameters.[4] The model was trained on a database of ~5000 lightcones of $140 \times 140 \times 2350$ pix at 1.4Mpc resolution. The training took only ~30 epochs (20 min/epoch with a moderate K80 GPU).

The 3D-21cmPIE-Net was able to jointly constrain cosmology, CD and EoR astrophysics without degeneracies and with small scatter and low bias for most parameters. In detail, L_X was estimated with 1σ errors $<1\%$, Ω_m and T_{vir} of few %, ζ down to few % depending on centre value, E_0 at the 10% level and m_{WDM} down to roughly 10% (considerably higher for $m_{WDM} \gtrsim 2$ keV). Interestingly, for all parameters except m_{WDM} the network's inference was insensitive to increasing

[1] Which here are: Typical X-ray brightness L_X, energy threshold for self-absorption in galaxies E_0, ionising efficiency ζ, minimum virial temperature for efficient star formation T_{vir}, matter density Ω_m and warm dark matter mass m_{WDM}.

[2] ResNet = Residual Network, uses skip connections between residual blocks.

[3] LSTM = long-short term memory network, efficiently memorises different parts of a series for feedback connections. LSTM have been successfully employed for video and audio analysis [5].

[4] 3D 21 cm Parameter InfErence Network: https://github.com/stef-neu/3D-21cmPIE-Net.

levels of foregrounds (instrumental noise, 21 cm foreground wedge, horizon and beam effects). Even transfer learning, i.e. training on simulations and evaluating on mock data with noise, and vice versa, yielded low scatter and bias especially for the X-ray luminosity L_X (and $<1\%$ bias across parameter range), but also robust inference of ionising efficiency ζ. This gives interesting insights on the effect of wrongly simulating noise levels in training data. For example for Ω_m training on noiseless simulations and evaluating on mocks yields increased scatter, while the other way round creates bias towards low Ω_m values. Thus our 3D network can:

- efficiently infer both cosmology and astrophysics from full 21 cm lightcones,
- provide low scatter and bias estimates of a multi-parameter set,
- estimate parameters insensitive to increased noise levels (except for m_{WDM}),
- robustly transfer learn simulations ↔ mock observations for some parameters.

4 Discussion

The main message is that **3D networks for 3D data work best**. This was tested for both low-redshift and high-redshift 3D data cubes of the 21 cm signal that traces neutral hydrogen across cosmic times. For low-redshift simulated data in preparation of SKA-MID an HI source detection and characterisation task was performed best by a 3D U-Net and CNN variants as compared to 2D U-Nets, R-CNN and inception models. The 3D U-Net was able to provide 3D reconstructions of sources without preprocessing suitable for estimation of source properties such as HI flux, independent of brightness. For high-redshift simulated data in preparation of SKA-LOW during CD and EoR, an inference task of astrophysical and cosmological parameters was performed best by a 3D CNN as compared to 2D CNN, ResNet, and LSTM models. Remarkably, the quality of the estimate was mostly insensitive to noise (Fig. 2).

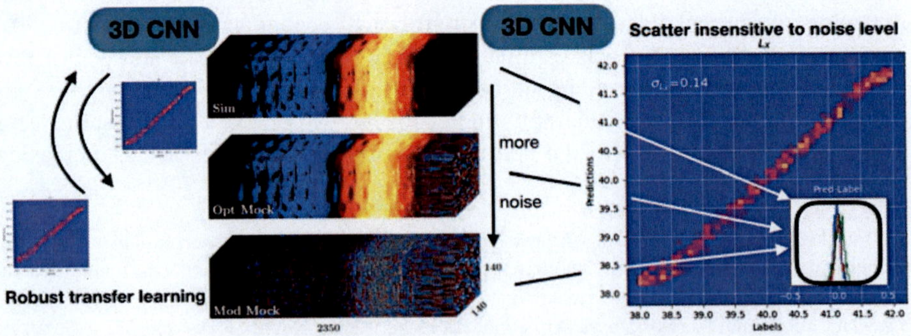

Fig. 2 Scheme of key findings for 3D CNN parameter inference from 21 cm lightcones [6]

Acknowledgments A special thanks and acknowledgement go to S. Neutsch as the main developer of the 3D-21cmPIE-Net, as well as to the team 'coin' of the SKA SDC2 and developers of our machine learning models, M. D. Veneri (INAF), B. Fraga (CBPF), A. Soroka (MSU), F. Gubanov (MSU), C. de Bom (CBPF), A. Meshcheryakov (IKI), and M. Brüggen (UHH).

References

1. Hartley, P., et al.: SKA science data challenge 2: analysis and results. Mon. Not. R. Astron. Soc. **523**(2), 1967–1993 (2023)
2. Ronneberger, O., Fischer, P., Brox, T.: U-Net: convolutional networks for biomedical image segmentation. In: MICCAI, pp. 234–241. Springer (2015). https://doi.org/10.48550/arXiv.1505.04597
3. Girshick, R., Donahue, J., Darrell, T., Malik, J.: Rich feature hierarchies for accurate object detection and semantic segmentation. In: CVPR (2014). https://doi.org/10.48550/arXiv.1311.2524
4. Szegedy, C., et al.: Going deeper with convolutions. In: ILSVRC (2014). https://doi.org/10.48550/arXiv.1409.4842
5. Hochreiter, S., Schmidhuber, J.: LSTM. Neural Comput. **9**(8), 1735–1780 (1997)
6. Neutsch, S., Heneka, C., Brüggen, M.: Inferring astrophysics and dark matter properties from 21 cm tomography using deep learning. Mon. Not. R. Astron. Soc. **511**(3), 3446–3462 (2022)

ConvNets for Enhanced Background Discrimination in the Diffuse Supernova Neutrino Background Search

D. Maksimović, M. Nieslony, and M. Wurm

Abstract The present paper describes a novel machine learning based method for the discrimination of a Neutral-Current (NC) background that taints IBD signals of the Diffuse Supernova Neutrino Background (DSNB) in Gadolinium-loaded large water Cherenkov detectors. Convolutional Neural Networks (CNNs) offer the possibility for a direct analysis and classification of photomultiplier (PMT) hit patterns of the prompt events in large water Cherenkov detectors. Based on the events generated in a simplified SuperKamiokande-like detector setup, we find that a trained CNN performs exceptionally well with a signal efficiency for the Diffuse Supernova Neutrino Background (DSNB) of 96% while reducing the residual NC background to 2% of the original rate. Compared to recent predictions of the DSNB signal and measurements of the NC background levels in Super-Kamiokande, the corresponding signal-to-background ratio is about 4:1, providing excellent conditions for a DSNB discovery.

1 Introduction

The first detection of the Diffuse Supernova Neutrino Background (DSNB) will be a milestone for low-energy neutrino astronomy. Despite the greatly improved conditions through Gadolinium doping of large scale water Cherenkov detectors like in the case of Super-Kamiokande (SK) for its current Phase VI run, DSNB detection is still beset by important backgrounds. IBDs induced by reactor and atmospheric $\bar{\nu}_e$'s define a narrow detection window from roughly 12 to 30 MeV. Within this observation window, the most prominent background is posed by Neutral-Current (NC) interactions of atmospheric neutrinos on oxygen in the water that are able to mimic the IBD coincidence signature. We demonstrate that the discrimination power of a Monte Carlo-trained CNN is sufficient to isolate the DSNB signal with high

D. Maksimović (✉) · M. Nieslony · M. Wurm
Institute of Physics and Excellence Cluster PRISMA+, Johannes Gutenberg-Universität Mainz, Mainz, Germany
e-mail: damaksim@uni-mainz.de

F. Bufano et al. (eds.), *Machine Learning for Astrophysics*, Astrophysics and Space Science Proceedings 60, https://doi.org/10.1007/978-3-031-34167-0_35

efficiency, achieving a signal-to-background ratio far greater than 1. This research is described in detail in the publication Maksimović et al. [5] and in short in this following paper.

2 Diffuse Supernova Neutrino Background

Massive stars ($M > 8\ M_{\odot}$) undergo a core-collapse supernova (CCSNe) at the end of their burning phase. The released gravitational binding energy caused by the gravitational collapse of the star cannot be successfully liberated in the dense core regions by photons, but only through neutrinos due to their weakly interacting nature. The DSNB is a proposed background composed of neutrinos (and anti-neutrinos) emitted by all the CCSNe throughout the Universe. The direct study and detection of emerging neutrinos in CCSNe can give insight into the condition of the collapsing core and the explosion mechanism, revealing the versatile role they play, whereas detecting the DSNB can also reveal additionally the fraction of hidden/failed SN events and an average Supernova neutrino-spectrum. Current estimates regarding the number of emitted neutrinos during a supernova explosion amount to around 10^{58} particles with equal partition between flavours due to eigenstates mixing through pair production during the proto-neutron star cooling [7].

A prime method for detecting the DSNB is finding the Inverse Beta Decay (IBD) signatures of the associated antineutrino signal in large Gadolinium-loaded water Cherenkov detectors like Super-Kamiokande (SK-Gd). While the enhanced neutron tagging capability of Gadolinium greatly reduces single-event backgrounds, correlated events mimicking the IBD coincidence signature remain a potentially harmful background. Especially in the low-energy range of the observation window, Neutral-Current (NC) interactions of atmospheric neutrinos dominate the DSNB signal, which leads to an initial signal-to-background (S:B) ratio inside the observation window of about 1:10, as presented in Fig. 1.

3 Detection and Background

While not yet observed, the DSNB signal can be predicted based on our current understanding of CCSNe and cosmology. We use the estimate for the DSNB flux estimated by Kresse et al. [4] which results in a flux in the order of $r \sim 10^2\ \mathrm{cm^{-2}s^{-1}}$ for an observer located on earth. In SK-Gd, the DSNB signal is dominated by the reactor neutrino background at visible energies below 12 MeV [2]. In the region of interest (12–30 MeV), the fiducial model predicts ~20 DSNB events in 10 years for a fiducial volume of 22.5 kt and a delayed neutron detection efficiency of 67%.

IBDs induced by reactor and atmospheric electron antineutrinos define borders for a narrow visible energy detection window from roughly 12 to 30 MeV. Within

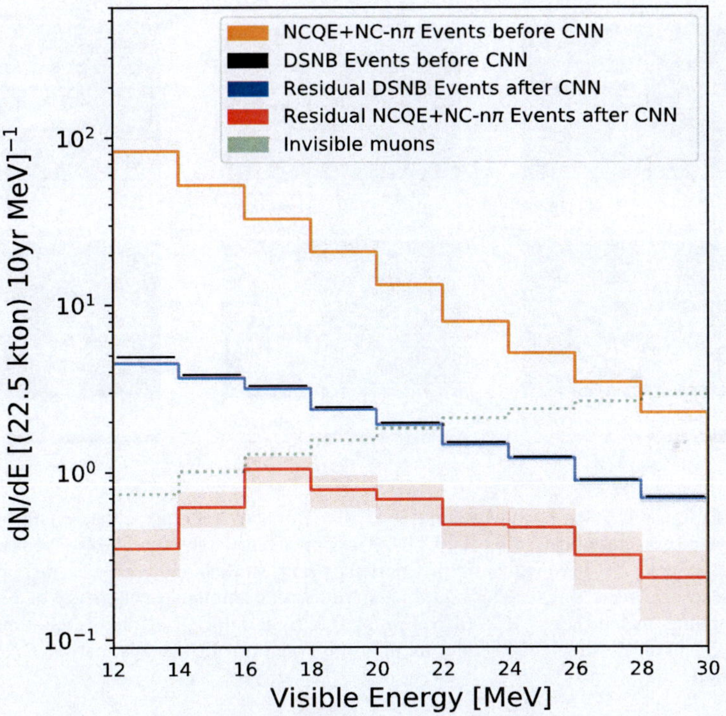

Fig. 1 This figure is taken from Maksimović et al. [5]. It shows the expected event rates for DSNB signal (*blue, black lines*) and NC background (*orange, red lines*) before and after the application of the CNN classification. A strict threshold value of $r \geq 0.965$ for the CNN was used. For comparison, we show the spectral contributions from invisible muons (*green dotted line*) that won't be affected by CNN classification [6]

this observation window, the most prominent background is posed by Neutral-Current Quasi-Elastic (NCQE) interactions of atmospheric neutrinos on oxygen in the water that are able to mimic the IBD coincidence signature. The coincidence signature is created by emitting one or several gammas and one or more neutrons in nuclear breakup reactions. While the hadronic fragments of an NC interactions are mostly below Cherenkov threshold, the gammas produced in de-excitation processes of the remnant nucleus are identified as the prompt event within the expected visible energy window.

Figure 2 displays the corresponding hit maps for event charge and timing for two exemplary events of DSNB and NCQE categories. To the human eye, the difference between a single strong Cherenkov ring from an IBD positron and several weak rings created by several low-energy gammas in the NCQE event is immediately apparent. In anticipation of the later section, the two right-most panels show so-called saliency maps that visualize the emphasis given by the trained CNN to specific regions of the 2D projections in the classification of these events. While

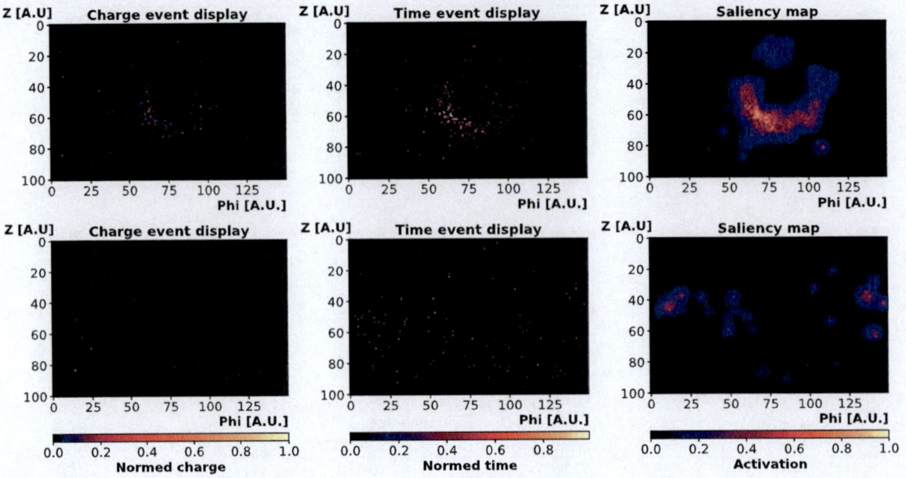

Fig. 2 This figure is taken from Maksimović et al. [5]. It shows a representative pair of event hit maps prepared for classification by the CNN. The upper panels show a DSNB event ($E_{\text{visible}} = 22\,\text{MeV}$, $E_{\text{true},e^+} = 25\,\text{MeV}$) with a single Cherenkov ring, while the lower panels display a NCQE event with several weak rings from multiple gammas with a summed total energy of 23 MeV and a corresponding visible energy of 19 MeV. From left to right, normalized charge and time hit maps and saliency maps are shown. The latter highlight the regions of the maps that the CNN uses for classification

the attention of the network is centered on the Cerenkov ring in case of the DSNB events, it is quite spread out for the NCQE map.

4 Convolutional Neural Network

The simulations of the DSNB signal and atmospheric NCQE background event samples are performed in `WCSim`, an open-source `Geant4`-based framework for modeling particle interactions and the corresponding detector response in Water Cherenkov detectors [1, 3]. The simulated DSNB and NCQE events are converted into pixelized images called "hit maps", which are conceptually close to conventional event displays. The collected charge (photo electrons, p.e.) per PMT and first photon arrival time are translated into two hit maps in which each PMT is assorted to a specific pixel. Pixels in both hit maps are normalized to a range from 0 to 1. In the case of charge information, the normalization is done relative to the largest p.e. number detected in the present event. This prevents the CNN from picking up on the absolute scale of the observed charge signal and thus largely prevents an energy bias in the classification. The CNN receives two-channel images with one charge channel and time channel in order to perform its classification.

To build and train the CNN, the Keras libraries and the Tensorflow framework for Python was used. The CNN features a total of 726k trainable and 1.2k non-trainable parameters and consists of six consecutive "64-double convolutional packages". Each package consists of two successive convolutional layers with 64 kernels each, followed by a 2-dimensional (2,2)-Maxpooling layer, a BatchNormalisation layer and a final Dropout layer in four of the six packages which is set to 20%. The first convolutional package has kernels with the dimension of (7,7), decreasing for the next packages from (5,5) to two times (3,3) and ending for the last two packages at (2,2). The classification result for the two categories DSNB and NCQE is One-Hot encoded in a two-dimensional target vector.

We use three different class-balanced and shuffled sets: a training sample, a validation sample and a final test sample. The training sample consists of 185,000 class-balanced events with 92,500 positron events and 92,500 NCQE events. The validation sample contains 15,000 events and is used to monitor the progression of the training and indicate possible adjustments of hyperparameters. During the training, we assure that the CNN is not overtrained by applying tensorflow callbacks such as *Modelcheckpoint* and monitor the performance of the validation set for each epoch. Our final test sample is composed of 20,000 events between 10 and 30 MeV (14,000 in the range 12–30 MeV) and is used only once for the evaluation of the CNN performance.

5 Conclusion and Results

By scaling the statistics of our DSNB and NC test samples to the numbers of signal and background events expected for a measuring time of 10 years in our simplified SK-Gd-like detector, one gets the initial signal-to-background (S:B) ratio of about 1:11 inside the detection window (12–30 MeV). By implementing and tuning the default discrimination threshold from $r \geq 0.5$ to $r \geq 0.965$ of our CNN, we reach 98% signal efficiency and 96% background rejection. After classification, the DSNB signal exceeds the residual atmospheric background level at a ratio of 4:1. In this, the discrimination power relies mostly on the inherent differences in event topologies (single strong vs. multiple washed-out rings) and thus primarily depends on the granularity of the light readout (i.e. the number of PMTs per photosensitive area). The behavior of the neural network has been rigorously probed and tested. That entire process is documented in Maksimović et al. [5].

In summary, CNN-based classification provides a real and systematically largely independent alternative to conventional methods of event selection for DSNB searches in large water-Gd detectors.

References

1. Agostinelli, S., et al.: GEANT4: a simulation toolkit. Nucl. Instrum. Meth. **A506**, 250–303 (2003)
2. Beacom, J.F., Vagins, M.R.: GADZOOKS! Anti-neutrino spectroscopy with large water Cherenkov detectors. Phys. Rev. Lett. **93**, 171101 (2004). https://doi.org/10.1103/PhysRevLett.93.171101
3. Himmel, A., O'Sullivan, E., Richards, B., Viren, B., Walter, C.: The Water Cherenkov Simulator (WCSim). https://github.com/WCSim/WCSim (2020)
4. Kresse, D., Ertl, T., Janka, H.T.: Stellar collapse diversity and the diffuse supernova neutrino background. Astrophys. J. **909**(2), 169 (2021). https://doi.org/10.3847/1538-4357/abd54e. Garching Core-Collapse Supernova Archive: https://wwwmpa.mpa-garching.mpg.de/ccsnarchive/archive.html
5. Maksimović, D., Nieslony, M., Wurm, M.: CNNs for enhanced background discrimination in DSNB searches in large-scale water-Gd detectors. J. Cosmol. Astroparticle Phys. **2021**(11), 051 (2021). https://doi.org/10.1088/1475-7516/2021/11/051
6. Møller, K., Suliga, A.M., Tamborra, I., Denton, P.B.: Measuring the supernova unknowns at the next-generation neutrino telescopes through the diffuse neutrino background. J. Cosmol. Astroparticle Phys. **2018**(05), 066–066 (2018). https://doi.org/10.1088/1475-7516/2018/05/066
7. Yoshida, T., Kajino, T., Yokomakura, H., Kimura, K., Takamura, A., Hartmann, D.H.: Neutrino oscillation effects on supernova light-element synthesis. Astrophys. J. **649**(1), 319 (2006). https://doi.org/10.1086/506374

Deep Neural Networks for Single-Line Event Direction Reconstruction in ANTARES

Juan García-Méndez, Miguel Ardid, and Salva Ardid on behalf of the ANTARES Collaboration

Abstract ANTARES, the first large undersea neutrino telescope, consisted of 12 vertical lines forming a 3D array of photo-sensors that detected Cherenkov light induced by secondary particles from neutrino interactions. Here, we present a method based on deep learning aimed to improve the direction reconstruction of low-energy single-line (SL) events in ANTARES, for which the reconstruction of the azimuthal angle of the incoming neutrino is particularly difficult. The architecture of the neural network combines the properties of Deep Convolutional Networks (DCNs) with those of Mixture Density Networks (MDNs). Results of the network on the test set showed a mean deviation upon the true value of 7.4° for the reconstruction of the zenithal angle, in contrast to 15.5° using former methods. We also computed the estimation of the azimuth, which was missing previously for SL events due to low precision. After several complementary tests, such as a fivefold cross-validation and the comparison with real data, we conclude that this method is ready to be applied in physics analyses.

J. García-Méndez
Universitat Politècnica de València, Research Institute for Integrated Management of Coastal Areas, Gandia, Spain

Design, Manufacture and Management of Industrial Projects, Universitat Politècnica de València, Valencia, Spain

M. Ardid · S. Ardid (✉)
Universitat Politècnica de València, Research Institute for Integrated Management of Coastal Areas, Gandia, Spain

Department of Applied Physics, Universitat Politècnica de València, Valencia, Spain
e-mail: sardid@upv.es

F. Bufano et al. (eds.), *Machine Learning for Astrophysics*, Astrophysics and Space Science Proceedings 60, https://doi.org/10.1007/978-3-031-34167-0_36

1 Introduction

ANTARES was the first undersea neutrino telescope [1]. It was composed of 12 vertical flexible lines anchored to the seabed. Each line was divided by 25 storeys, containing photomultipliers (PMTs) inside glass spheres called Optical Modules (OMs). In each storey there were three OMs, separated 120° in the horizontal plane and looking downwards at 45°.

ANTARES recorded events *run-by-run*, where a run is the period in which the telescope takes data with a given configuration, typically for few hours. The information of a recorded event includes the run number, the position of OMs, their direction, the environmental conditions and the information of the PMTs in terms of hits.

The principal reconstruction method for SL events in ANTARES is based on a χ^2 fit [2], which does not predict the azimuth of the incoming neutrino. Thus, we propose a deep learning method aimed to improve the reconstruction of the zenith (θ) and a first estimation of the azimuth (ϕ).

2 Deep Networks Development and Training

The neural networks developed in this work are a combination of Deep Convolutional Networks (DCNs) [3] and Mixture Density Networks (MDNs) [4]. Many architectures and network parameters were explored. The result of this exploration determined that two separate networks for the θ and ϕ angles worked best.

To predict not only the reconstructed values (μ) but also their uncertainties (σ), we use a MDN output layer. This requires replacing the loss function ($\mathcal{L}$), the mean square error typically used in point predictions, by the negative log-likelihood. For this, we assume a normal distribution, so that the resulting loss is

$$\mathcal{L} = \ln\left(\sqrt{2\pi}\sigma\right) + \frac{1}{2}\left(\frac{y_t - \mu^2}{\sigma}\right), \quad (1)$$

where y_t refers to the true value. For further information on the architecture, see [5].

The ANTARES collaboration has produced Monte Carlo (MC) simulations [6] that take into account many variables such as environmental variations, the telescope conditions or the setup, which are well suited to apply supervised learning algorithms. Our study focuses on track-like events originated from charged current interactions of muon-(anti)neutrinos (ν_μ^{CC}), selecting SL events according to the goodness of the χ^2 fit.

We transformed the information of each event into an image that we then fed to the DCN. Since we focus only in SL events, we create 2D images, with time in the x-axis and the telescope's vertical storey in the y-axis (see [5] for further details).

3 Results and Conclusions

Statistics about the absolute difference between the reconstructed and actual angles are reported in Table 1. Distribution plots can be seen in [5]. Our DCN model outperforms the method based on the χ^2 fit by showing a larger proportion of small for the zenith (see [5]), and a first estimation of the azimuth, which was previously missing for SL events.

To ensure that our method is ready to use for real physics analyses, we perform additional tests. Two five-fold cross-validation analyses were performed, one with randomly split sets and the other according to date. The two analyses showed stable errors comparable to those in Table 1. We also checked that these results can be applied to real analyses by comparing the distributions of the predicted angles of real data and MC simulations. This can be seen in Fig. 1. Thus, we conclude that the approach has proven to be robust and that it is ready to be applied to actual physics analyses.

Table 1 Statistics of the difference (in absolute values) between the reconstructed angles and the true neutrino direction

	BBfit		DCN		DCN (50% lowest σ)	
	Mean	Median	Mean	Median	Mean	Median
θ	15.5°	8.5°	7.4°	4.4°	3.7°	2.5°
ϕ	–	–	41.4°	31.5°	29.2°	23.6°
Ω	–	–	28.3°	22.7°	18.7°	13.7°

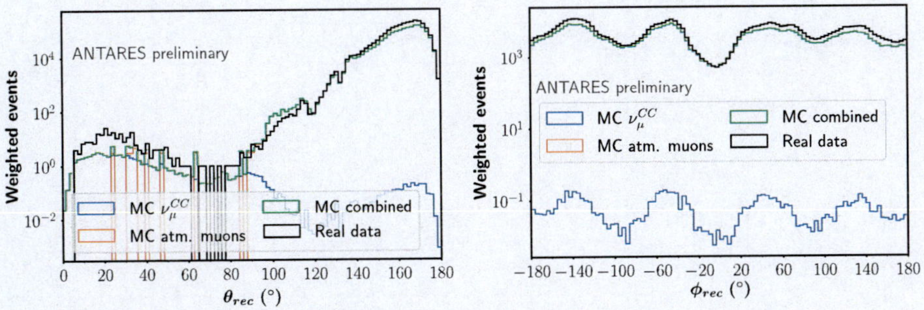

Fig. 1 Comparison between real and MC data using quality cuts (σ_θ<5°, σ_ϕ<20°)

Acknowledgments Generalitat Valenciana Gen-T Program (ref. CIDEGENT/2019/043), Ministerio de Ciencia e Innovación/European Union (FEDER): Programa Estatal de Generación de Conocimiento (ref. PGC2018-096663-B-C43 and PID2021-124591NB-C42), and "European Union NextGenerationEU/PRTR", Programa de Planes Complementarios I+D+I (ref. ASFAE/2022/014). We also acknowledge the computer resources at Artemisa, funded by the European Union ERDF and Comunitat Valenciana.

References

1. ANTARES collaboration: ANTARES: The first undersea neutrino telescope. Nucl. Instrum. Meth. A **656**, 11–18 (2011)
2. ANTARES collaboration: A fast algorithm for muon track reconstruction and its application to the ANTARES neutrino telescope. Astropart. Phys. **34**, 652–662 (2011)
3. Gu, J., et al.: Recent advances in convolutional neural networks. Pattern Recogn. **77**, 354–377 (2018)
4. Bishop, M.: Mixture density networks. NCRG/94/004 (1994)
5. García-Méndez, J., et al.: Deep learning reconstruction in ANTARES. J. Instrum. **16**, C09018 (2021)
6. ANTARES collaboration: Monte Carlo simulations for the ANTARES underwater neutrino telescope. J. Cosmol. Astropart. Phys. **1**, 064 (2021)

Cats vs Dogs, Photons vs Hadrons

Francesco Visconti

Abstract In gamma ray astronomy with Cherenkov telescopes, machine learning models are needed to guess what kind of particles generated the detected light, and their energies and directions. The focus in this work is on the classification task, training a simple convolutional neural network suitable for binary classification (as it could be a *cats vs dogs* classification problem), using as input uncleaned images generated by Montecarlo data for a single **ASTRI** telescope. Results show an enhanced discriminant power with respect to classical random forest methods.

1 The Context: Detect Cherenkov Light with ASTRI Telescopes

Ground-based gamma-ray astronomy is a young field with enormous scientific potential. The current generation instruments *H.E.S.S.*,[1] *MAGIC*[2] and *VERITAS*[3] have already demonstrated the huge physical potential of astrophysical measurements at teraElectronvolt (TeV) energies. The 250+ sources detected by these instruments (http://tevcat.uchicago.edu/), and the wide range of high impact scientific results, suggest that particle acceleration yielding TeV gamma-rays is common in nature.

[1] https://www.mpi-hd.mpg.de/hfm/HESS/HESS.shtml.

[2] https://magic.mpp.mpg.de/.

[3] https://veritas.sao.arizona.edu/.

F. Visconti (✉)
Osservatorio Astronomico di Roma, INAF, Monte Porzio Catone, RM, Italy
e-mail: francesco.visconti@inaf.it

F. Bufano et al. (eds.), *Machine Learning for Astrophysics*, Astrophysics and Space Science Proceedings 60, https://doi.org/10.1007/978-3-031-34167-0_37

The *ASTRI* [1] (Astrofisica con Specchi a Tecnologia Replicante Italiana) telescope is an INAF project falling in the category of Cherenkov Telescope Array's Small Size Telescope (SST) with a diameter of 4 m, with double mirror technology. The ASTRI's data reduction software [2] has been entirely developed by a dedicated INAF group of work. This work is focused on the reconstruction stage, and in particular in the piece of software needed to discriminate between real signal and background.

2 Network and Data

Any data analysis software for Cherenkov telescopes needs a fundamental component to discriminate if the air shower has been produced by photons or cosmic rays; this is done with machine learning techniques, training models on simulated data (Fig. 1).

The gold standard in this field is Random Forest [3], here implemented with `scikit-learn` library [4] which is run over Hillas parameters: this method introduces a bias (Hillas parameters are calculated after image cleaning and assuming a shape of the Cherenkov light in the image), which can be overcome by deep learning techniques [5], here implemented with the `keras` [6] library.

With that aim, a five layers Convolutional Neural Network has been trained with ASTRI calibrated images from a Montecarlo production. For input data, single telescope images have been generated starting from a Montecarlo ASTRI data production, calibrated via the dedicated software developed in Rome by our group [2]. Events data are selected before the cleaning process, keeping the full background of the pixels due to the Night Sky Background (NSB). Those images

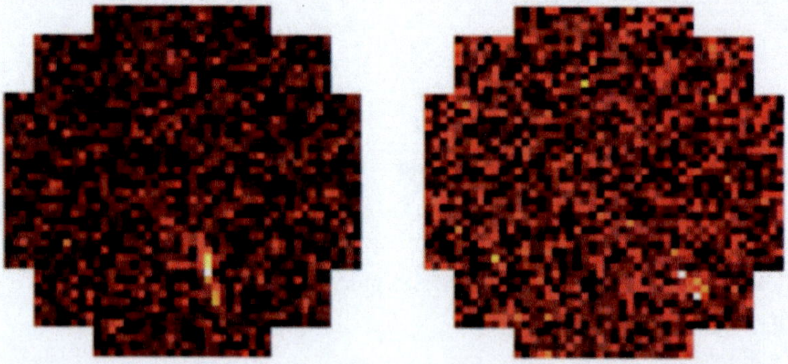

(a) (b)

Fig. 1 An example of calibrated Cherenkov events as seen by ASTRI telescope. (**a**) Gamma. (**b**) Proton

Table 1 Convolutional Neural Network architecture used for the classification task

Layer (type)	Output shape	n. of Params
Conv2D	(None, 54, 54, 32)	896
MaxPooling2D	(None, 27, 27, 32)	0
Conv2D	(None, 25, 25, 32)	9248
MaxPooling2D	(None, 12, 12, 32)	0
Conv2D	(None, 10, 10, 64)	18,496
MaxPooling2D	(None, 5, 5, 64)	0
Dense	(None, 64)	102,464
Dropout	(None, 64)	0
Dense	(None, 1)	65
Total (all trainable) params: 131,169		

are the signal received on any given camera pixel, a way different input from Hillas parameters (tabular data) characterizing the events. 70k images have been used for each class in the training step, 30k for validation during training, and finally prediction are made on 100k files for each class.

In Table 1, the network architecture adopted for the task is shown.

3 Results

The training took 1 hour on a K20 NVIDIA GPU, stopped after 11 epochs monitoring the validation accuracy improvement. With regard to the performance, the max validation accuracy reached was 96.7%. For a quick comparison, the standard method employed in the ASTRI Mini Array data analysis and reconstruction (Random Forest on Hillas parameters) do not exceed 86% in our case. The good result is confirmed by the gamma/hadron separation plot,[4] here plotted in log-scale to better appreciate the distributions, where we can see an enhanced discrimination power for the CNN network on calibrated data. All other metrics taken into account (ROC curve with AUC score, Precision vs Recall, Confusion Matrix) confirm that **CNN performs significantly better than RF**.

In Table 2, other useful metrics[5] are summarized, all pointing in the same direction: this CNN outperforms the Random Forest.

[4] This kind of plot is an histogram where for each probability bin we count (actually there's a density on the y axis) the number of particles classified as gamma with that probability: the more the two populations are separated, the better.

[5] For a list of useful metrics see for example https://scikit-learn.org/stable/modules/model_evaluation.html.

Table 2 Classification metrics comparison between Random Forest and CNN

	Val Acc	AUC	f1	brier loss	TP	FN	FP	TN
Random Forest	86%	0.94	0.92	0.09	84%	16%	11%	89%
Convolutional NN	**96.7%**	**0.99**	**0.97**	**0.03**	**96%**	**4%**	**3%**	**97%**

4 Summary

The work produced interesting results with a small neural network, enhancing the discriminatory power for the gamma-hadron separation problem. This encourages to explore these techniques, suggesting the bias introduced by event parametrization is very important. Future improvements are currently in progress using "stereo" images (events seen by more than one telescope), with a pipeline able not only to do signal-background classification but also energy and direction reconstruction [7].

References

1. Scuderi, S., et al: The ASTRI mini-array of Cherenkov telescopes at the Observatorio del Teide. J. High Energy Astrophys. **35**, 52–68 (2022). https://doi.org/10.1016/j.jheap.2022.05.001
2. Lombardi, S., et al: Very-high-energy astrophysics, imaging atmospheric Cherenkov telescopes, CTA, ASTRI, data reconstruction and analysis methods, pipelines, Monte Carlo simulations. Proc. SPIE, 10707 (2018). https://doi.org/10.1117/12.2311293
3. Breiman, L.: Random Forests. Mach. Learn. **45**, 5–32 (2001). https://doi.org/10.1023/A:1010933404324
4. Pedrosa, F., et al.: Scikit-learn: machine learning in Python. J. Mach. Learn. Res. **12**, 2825–2830 (2011)
5. Shilon, I., et al: Application of deep learning methods to analysis of imaging atmospheric Cherenkov telescopes data. Astropart. Phys. **105**, 44–53 (2019). https://doi.org/10.1016/j.astropartphys.2018.10.003
6. Chollet, F., et al.: Keras. https://keras.io
7. Miener, T., et al.: Reconstruction of stereoscopic CTA events using deep learning with CTLearn. In: PoS, ICRC2021 (2021). https://doi.org/10.22323/1.395.0730

Events Classification in MAGIC Through Convolutional Neural Network Trained with Images of Observed Gamma-Ray Events

S. Truzzi, G. Bonnoli, R. Cappuccio, R. Paoletti, T. Miener, and on behalf of the MAGIC Collaboration

Abstract Imaging Atmospheric Cherenkov telescopes, such as the Major Atmospheric Gamma Imaging Cherenkov (MAGIC) telescope, are able to detect from the ground gamma rays with energies beyond several tens of GeV, emitted by the most energetic celestial sources. Events classification is needed to classify gamma rays among the dominant cosmic rays. In the MAGIC standard analysis this task is performed by means of a Random Forest, a machine learning algorithm that works on a parameterization of the stereoscopic images.

Convolutional neural networks are a type of artificial neural network designed to exploit the whole image information. Other works in this field perform the network training on Monte Carlo simulated data. This choice is prone to systematics arising from discrepancies between observational data and simulations. Instead, in our work, we trained for the first time a CNN-based model (Inception Resnet V2) with observational data from a giant flare of the Markarian 421 blazar observed by MAGIC in 2013. Using this method for gamma/hadron separation, the preliminary results compete with the standard MAGIC analysis.

S. Truzzi (✉) · R. Cappuccio · R. Paoletti
University of Siena, Department of Physical Sciences, Earth and Environment, Siena, Italy

INFN Pisa, Pisa, Italy

G. Bonnoli
INAF Osservatorio Astronomico di Brera, Milano, Italy

INFN Pisa, Pisa, Italy

T. Miener
IPARCOS Institute, Madrid, Spain

EMFTEL Department, Universidad Complutense de Madrid, Madrid, Spain

F. Bufano et al. (eds.), *Machine Learning for Astrophysics*, Astrophysics and Space Science Proceedings 60, https://doi.org/10.1007/978-3-031-34167-0_38

1 Gamma/Hadron Classification

The MAGIC telescopes [1] have been designed to reach the lowest possible energy threshold by detecting the faintest signal of extended air showers generated in the atmosphere by energetic primary particles of celestial origin. The air showers generate optical flashes through Cherenkov effect and are imaged at the nanosecond scale by the telescopes.

In the MAGIC standard analysis (SA), event information, such as charge and arrival time distributions, is used for reconstruct the incident gamma ray physical properties. The former is translated into a set of geometrical features known as Hillas parameters [2]. This approach was adopted when the hardware resources were limited. Current computer technology, especially the parallel computing capability of the Graphic Processing Unit (GPU), can easily and quickly process large volumes of data, allowing more refined algorithms for Imaging Atmospheric Cherenkov Telescopes (IACT) events reconstruction.

In our work we exploit the Convolutional Neural Network (CNN) scheme Inception Resnet V2 (IRV2) [3] to classify the MAGIC data in the two classes of "hadronic" and "gamma-ray" events, by training the CNN with images of real events. This choice potentially avoids systematic effects introduced by discrepancies between simulated and observed events [4]. Adding Poisson noise to the simulated images in order to match the Night Sky Background [5] or cleaning the images [6] have been proposed as alternative ways to overcome this issue. Here we explore for the first time the potential of training a machine learning algorithm with real IACT data only.

2 Training the CNN

For the training of the CNN we chose the Mrk421 extraordinary flare of April 10th–16th, 2013. This data sample has been extensively studied by Acciari et al. [7].

Following a similar procedure, we calculated the shower arrival direction and the hadronness (h^1). Our procedure included the following steps:

1. Conversion of the ROOT data into Python data with `Uproot` package.
2. Linking of hadronness and arrival direction to the original images.
3. Application of "absolute cleaning" [8] with $Q_{core} = 8$ and $Q_{boundary} = 4$.
4. Conversion of the hexagonal-grid images to square-grid images using the `DL1-DATA-HANDLER` package for adaptation to Tensorflow [9].

[1] Hadronness is a number between 0 and 1 associated to each event by the Random Forest classification implemented in the SA and providing an estimation of how likely is an hadronic origin for that event.

5. Definition of the training data set by circular regions in the camera with an aperture $\theta^2 = 0.02$. The gamma-like data set is centered on the source direction (ON region) and the hadron-like data sets are placed in three symmetrical locations (OFF regions); all positions are offset by $w = 0.4$ from the camera center [10].
6. Event selection based on the hadronness: $h \leq 0.5$ on gamma-like and $h \geq 0.5$ on hadron-like data sets.
7. Augmentation of the images via rotation/flip during training to have higher statistics since the original selection was based on few positions in the camera.

After these operations we took the CNN scheme known as IRV2 and adapted the top layer to the image size. At the end of the training we obtained: $ValidationLoss = 0.173$ and $accuracy = 0.943$.

3 Performance of the CNN and Cross-Checks

We evaluated the performance of the CNN on another flare of Mrk421 which happened on March 30th and April 1st, 2nd, 2013. Figure 1 on the left shows that the Excess(=ON-OFF) for the same background level (OFF) are slightly better for CNN than SA. On the right, the performances of CNN and SA are shown, confirming that the CNN analysis results are comparable or better than the SA ones.

As a control sample, we took the observations of the 1ES0927+500 blazar [11], where no gamma-ray excess has been detected by MAGIC, in order to check that the CNN does not artificially misinterpret background events as gamma-like events in a systematic way. The θ^2 distribution of Fig. 2 (left) shows similar behaviour for the ON and the three OFFs regions. Therefore, the CNN does not detect any significant

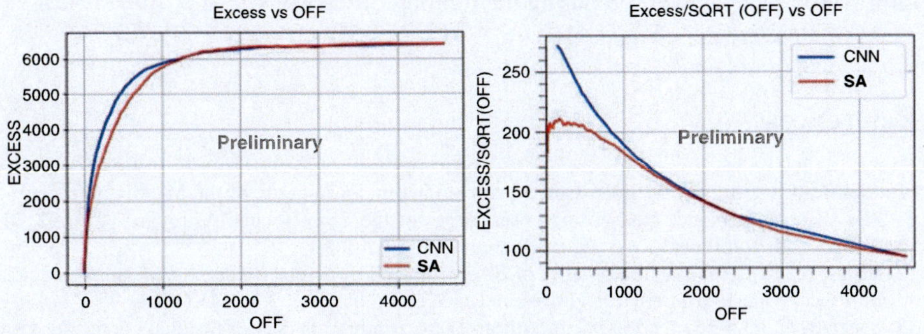

Fig. 1 The results obtained with IRV2 on a Mrk421 flare event sample, compared to the SA. An "8-4" absolute cleaning was applied. The gamma-ray $Excess$ above the background (left) and a measure of its statistical significance $Excess/SQRT(OFF)$ (right) are plotted as a function of the level of background events (OFF) surviving the hadronness cuts

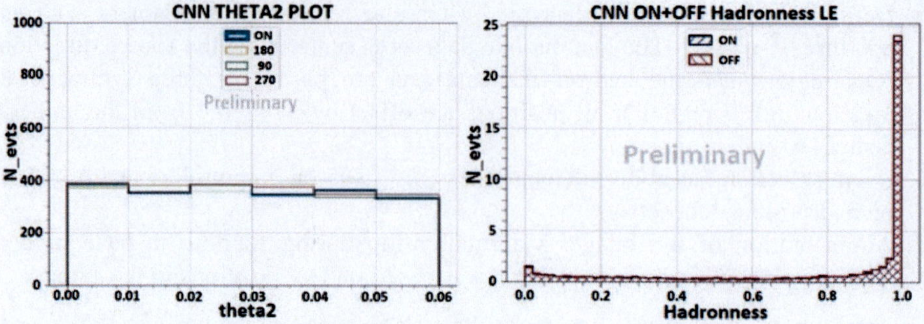

Fig. 2 Left: θ^2 plot for 1ES0927+500 showing similar trend for the CNN analysis on ON and OFF samples. Right: normalized hadronness CNN distributions for 1ES0927+500 for the ON (blue) and OFF (red) samples

gamma signal from those observations. The hadronness distribution (right) for 1ES0927+500 shows that the CNN identified most of the ON events as hadronic.

4 Conclusion and Future Implication

In this work we have shown that this CNN approach provides a valid alternative to state-of-the-art IACT analysis for event classification. We used for the first time observational data to train CNN-based models, avoiding systematic effects potentially introduced by the use of simulated events. The method certainly has room for further improvements that we are currently exploiting, such as the use of unsupervised neural network algorithms to avoid the use of the SA hadronness or the addition of other parameters, such as the arrival time distribution, to add more information that can be used during the training of the CNN.

References

1. Aleksic, J., et al.: The MAGIC Collaboration: The major upgrade of the MAGIC telescopes, Part II: A performance study using observations of the Crab Nebula. Astropart. Phys. **72**, 61 (2016). https://doi.org/10.1016/j.astropartphys.2015.02.005
2. Hillas, A.M.: Cerenkov light images of EAS produced by primary gamma rays and by nuclei. In: Proc. of the 19th International Cosmic Ray Conference, vol. 3, p. 445 (1985)
3. Szegedy, C., et al.: Inception-v4, Inception-ResNet and the impact of residual connections on learning. arXiv (2016). https://doi.org/10.48550/arXiv.1602.07261
4. Shilon, I., et al.: Application of deep learning methods to analysis of imaging atmospheric Cherenkov telescopes data. Astropart. Phys. **105**, 44 (2019). https://doi.org/10.1016/j.astropartphys.2018.10.003

5. Jacquemont, M., et al.: First full-event reconstruction from imaging atmospheric Cherenkov telescope real data with deep learning. arXiv:2105.14927 (2021). https://doi.org/10.1109/CBMI50038.2021.9461918
6. Miener, T., et al.: IACT event analysis with the MAGIC telescopes using deep convolutional neural networks with CTLearn. arXiv:2112.01828 (2021). https://doi.org/10.48550/arXiv.2112.01828
7. Acciari, V.A., et al.: Unraveling the complex behavior of Mrk 421 with simultaneous X-ray and VHE observations during an extreme flaring activity in 2013 April. Astrophys. J. Suppl. Ser. **248**, 29 (2020). https://doi.org/10.3847/1538-4365/ab89b5
8. Aliu, E., et al.: Improving the performance of the single-dish Cherenkov telescope MAGIC through the use of signal timing. Astropart. Phys. **30**, 293 (2009). https://doi.org/10.1016/j.astropartphys.2008.10.003
9. Abadi, M., et al.: TensorFlow: large-scale machine learning on heterogeneous distributed systems. arXiv:1603.04467 (2016). https://doi.org/10.48550/arXiv.1603.04467
10. Palacio, J., et al.: Pointing optimization for IACTs on indirect dark matter searches. Astropart. Phys. **104**, 84 (2019). https://doi.org/10.1016/j.astropartphys.2018.09.002
11. Foffano, L., et al.: A new hard X-ray selected sample of extreme high-energy peaked BL Lac objects and their TeV gamma-ray properties. Mon. Not. R. Astron. Soc. **486**, 1741 (2019). https://doi.org/10.1093/mnras/stz812

Federated Learning Meets HPC and Cloud

Iacopo Colonnelli, Bruno Casella, Gianluca Mittone, Yasir Arfat, Barbara Cantalupo, Roberto Esposito, Alberto Riccardo Martinelli, Doriana Medić, and Marco Aldinucci

Abstract HPC and AI are fated to meet for several reasons. This article will discuss some of them and argue why this will happen through the methods and technologies underpinning cloud computing. As a paradigmatic example, we present a new Federated Learning (FL) system that collaboratively trains a deep learning model in different supercomputing centers. The system is based on the StreamFlow workflow manager designed for hybrid cloud-HPC infrastructures.

1 Introduction

High-Performance Computing (HPC) and Artificial Intelligence (AI) are fated to meet for many reasons. First, HPC infrastructures embrace GPUs for their superior performance-per-watt ratio against general-purpose multicores. Second, the next-generation scientific workflows integrate AI-based steps for their accuracy in approximating and analyzing complex phenomena. Third, AI, specifically Machine Learning (ML), is a perfect workload for GPUs for required performance and development time. Today we cannot close the circle of seamlessly running AI-enabled scientific workloads into HPC infrastructures. On one side, HPC system software and development tools are not designed for modern workloads. For example, queue-based resource managers and security limitations to outbound connections hinder distributed and federated cross-cluster workloads. On the other side, frameworks for modern workloads, such as ML frameworks, are typically designed primarily for the cloud ecosystem. HPC-cloud convergence is likely to bridge the gap.

This work takes Federated Learning (FL) as a paradigmatic example of modern AI applications. FL tackles the problem of collaboratively training an ML model using distributed data silos, where the data within silos cannot be exposed outside

I. Colonnelli (✉) · B. Casella · G. Mittone · Y. Arfat · B. Cantalupo · R. Esposito · A. R. Martinelli · D. Medić · M. Aldinucci
Università degli Studi di Torino, Computer Science Department, Turin, Italy
e-mail: iacopo.colonnelli@unito.it

F. Bufano et al. (eds.), *Machine Learning for Astrophysics*, Astrophysics and Space Science Proceedings 60, https://doi.org/10.1007/978-3-031-34167-0_39

the infrastructure (and the owner-tenant) for privacy and secrecy concerns. FL is challenging because it requires federating infrastructures and orchestrating their workload managers (e.g., SLURM, PBS for HPC) cyclically. Indeed, training an FL model requires repeating several (possibly synchronous) rounds while exchanging information among infrastructures at each round. Moreover, the traditional HPC toolkit does not adequately support the design of this kind of application, as the FL workflow needs to express cyclic workflows spawning across different infrastructures (OS/architecture). The following two sections recap how FL works (Sect. 2) and how hybrid workflows can be used to WMS describe and orchestrate cloud-HPC workloads (Sect. 3). Section 4 presents a hybrid workflow describing FL across different HPC clusters with a cloud-based aggregator.

2 Federated Learning

More and more companies are adopting AI models to support their daily decisions, thanks to the growing availability of big data and computing power. Data are often fragmented between different organizations, forming the so-called "data silos". (FL) is a learning paradigm where multiple *workers* collaborate in solving a learning task using their private data. Importantly, each worker's local data is not exchanged or transferred to any other participant.

Federated Learning gained popularity for its capability to extract information from inherently distributed parties while keeping the data segregated in each party. Beyond privacy implications, keeping the data local is beneficial for performance and data integrity. For this, FL can be an enabling methodology for federating distributed research infrastructures, such as the Square Kilometre Array (SKA) [1].

The typical runtime architecture of FL frameworks (e.g., Intel OpenFL [2] and Flower [3]) is a master/worker. Each worker is deployed onto a different location owning a private portion of the federated dataset, where it trains a private copy of a Deep Neural Network (DNN) model. At the end of each training round, each worker sends the serialized state of its model to the master, which computes an aggregated model using a configurable aggregation strategy and broadcasts it back to workers for the next round. The first FL aggregation algorithm, proposed in 2016 by Google, is `FedAvg` [4], which works with (DNNs). In this setting, the aggregation plane performs a weighted average of the resulting models at the end of each round. DNNs are still the most used models in FL, even if works training standard machine learning models in federated settings exist in literature [5].

The fact that data in different silos might not be assumed to be independent and identically distributed (IID) poses a pivotal challenge to FL [6]. In the last few years, several FL algorithms (e.g., `FedCurv` [7]) have been proposed to tackle the problem of non-IIDness, and a recent article [8] shows that tuning the number of local gradient descent steps might be vital to attacking this problem, as it empirically shows that often algorithms perform better when this number is increased. Some recent FL frameworks also drop the constraint of a single centralized aggregator, either relying on a tree-based infrastructure or implementing a fully decentralized peer-to-peer aggregation protocol. However, since the runtime infrastructure and the communication topology are always intrinsic characteristics of the framework implementation, users cannot seamlessly customize them.

3 Hybrid Workflows

A *hybrid workflow* is a workflow whose steps can span multiple, heterogeneous, and independent computing infrastructures. Support for *multiple* infrastructures implies that each step can target a different execution *location*, which must have access to all the input dependencies. Locations can be *heterogeneous*, exposing different hardware resources and protocols for authentication, communication, resource allocation, and job execution. Plus, they can be *independent* of each other, meaning that direct communication among them may not be allowed. A hybrid workflow model should then be *composite*, enclosing a *workflow graph* that describes a complex application in terms of steps and their dependencies, a *location topology* that represents a heterogeneous execution environment in terms of HW/SW resources and communication channels, and a *mapping relation* of each workflow step onto one or more execution locations. Each location can have two distinct roles. A *control* location can execute a command locally, offload it to other locations, or initiate a data transfer. Conversely, a *processing* location can only execute or delegate commands sent by other locations.

Hybrid workflows unify a description of software components' interactions (steps and dependencies) with a description of the execution environment (locations and channels) in a unique model while keeping the two planes well separated. The global knowledge of the execution environment allows a Workflow Management System (WMS) to act as an orchestrator, managing locations' lifecycles and data transfers automatically without requiring modifications at the infrastructure level. As a practical example, they offer a way to federate HPC infrastructures at the application level, managing them as accelerators [9], i.e., offloading specific portions of complex workflows running elsewhere, from the user's workstation to general-purpose Cloud VMs. In addition, the explicit mapping of steps onto location allows for avoiding unnecessary data transfers, making hybrid workflows suitable for describing distributed applications where data movements are not allowed, like FL workloads.

4 Hybrid Federated Learning Workflows

In research scenarios, cross-silo FL pipelines are prone to run on heterogeneous and independent execution environments. Indeed, data providers are usually independent entities with different data treatment protocols, storage infrastructures, and access policies. Unfortunately, this independence also leads to inconsistent data formats, which need ad-hoc preprocessing pipelines before joining an FL experiment. Moreover, data protection measures are likely to disallow direct inbound and outbound connections involving the data storage nodes. Conversely, existing FL frameworks are designed to work with homogeneous infrastructures by deploying an agent on each worker node or relying on Function-as-a-Service (FaaS) frameworks to submit training commands. Most of them also require direct bidirectional interconnections between the aggregation control plane and the training nodes to exchange models' states directly. Representing an FL pipeline as a hybrid workflow can solve these problems.

This work evaluates hybrid workflows to execute a cross-cluster FL pipeline, where two independent HPC clusters train two different models on a private dataset and a Cloud VM acts as a centralized aggregator (see Fig. 1). The pipeline trains a VGG16 [10] with Group Normalization [11] over two datasets: a standard MNIST [12], residing on the CINECA MARCONI100 HPC facility located in Bologna (node: 2 × 16-core IBM POWER9 AC922, 256 GB RAM and 4 NVIDIA V100 GPUs), and a grayscaled version of SVHN [13], residing within HPC4AI [14] facility located in Torino (node: 80-core Arm Neoverse-N1, 512 GB RAM, 2 NVidia A100 GPU). The same DNN and datasets have been tested as a baseline in an article [15] showing that the aggregation of parameters at test time is a novel form of Transfer Learning.

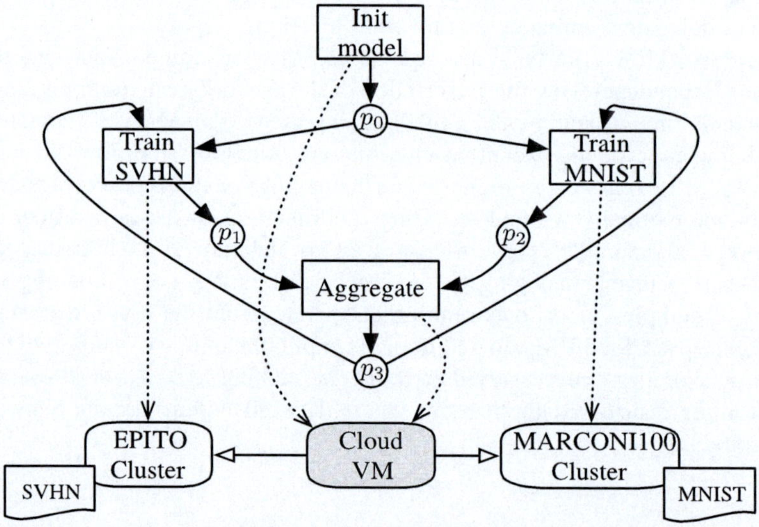

Fig. 1 Hybrid workflow model of an FL pipeline mapped onto a cloud+HPC topology. Steps are represented as squares, ports as circles, and dependencies as arrows with black-filled heads. Locations are represented as squashed rectangles, and communication channels as arrows with white-filled heads. Control locations are grey. Processing locations are blank. Mapping relations are expressed as dotted arrows

Table 1 Accuracy and time to solution for all the evaluated FL pipelines. OpenFL has not been evaluated on the Hybrid Cloud+HPC environment because it does not support this configuration, as it requires direct bidirectional communications between the aggregator and each worker node

		StreamFlow			OpenFL		
		MNIST acc.	SVHN acc.	Time	MNIST acc.	SVHN acc.	Time
Cloud	100 rounds, 1 epoch/round	99.36%	92.74%	2 h 40 m	97.91%	93.15%	3h06m
	50 rounds, 2 epochs/round	99.37%	92.74%	2 h 20 m	98.88%	94.21%	2h09m
Hybrid	100 rounds, 1 epoch/round	99.29%	93.06%	2 h 57 m	–	–	–
	50 rounds, 2 epochs/round	99.34%	92.85%	1 h 45 m	–	–	–

The FL pipeline has been described using an extension of the Common Workflow Language (CWL) open standard [16], a vendor-agnostic language based on declarative YAML files. The official CWL standard does not include iterative patterns, which are fundamental to expressing the FL training process. However, a `Loop` construct has recently been proposed as an extension.[1] As far as the authors know, the present work describes the first real-case iterative CWL workflow. At runtime, the StreamFlow framework [17] has been used to orchestrate the workflow over the hybrid cloud+HPC infrastructure described above. Two FL configurations have been tested: 100 rounds of 1 epoch each and 50 rounds of 2 epochs each. Both used the `FedAvg` aggregation strategy. In addition, to compare performances with a baseline, the same configurations have been tested on a pure cloud version of the pipeline, in which two worker VMs have executed the training steps (8 cores, 32 GB RAM, 1 NVIDIA T4 GPU each) using both Intel OpenFL [2] and StreamFlow (see Table 1). The two frameworks are comparable in terms of accuracy and time-to-solution when running on cloud VMs, meaning that general-purpose hybrid workflows are not "too general" to provide adequate performance. The overhead introduced by the remote cross-HPC executions (mainly due to queue times and transfers over low-performing networks) is largely compensated by the computing power and high-end interconnections provided by HPC facilities, and the benefits increase with heavier jobs (i.e., with more epochs per round, but also with larger datasets and models).

[1] https://github.com/common-workflow-language/common-workflow-language/issues/495.

5 Conclusion and Future Work

This work investigates the potential of hybrid workflow models to realize FL systems across heterogeneous infrastructures with limited connectivity between worker nodes (e.g., HPC modules). To the authors' knowledge, the reported experiment is the first usage of FL across supercomputers. Also, it suggests an effective way to federate HPC centers at the application level. The overhead introduced by using a general-purpose workflow system does not significantly hinder execution time. A drawback is the need to execute each round in each supercomputer in a single bulk. Indeed, an external coordination language like CWL cannot be used to separate different portions of the same Python script in different workflow steps. Moreover, by design CWL is unaware of the application internals and cannot deal with Python-specific dependencies (e.g., serialize and deserialize the program context). Some finer-grained alternatives will be explored in the future, like Jupyter-based literate hybrid workflows [18]. The code used for experimental evaluation is publicly available.[2]

Acknowledgments This work has been partially supported by the ACROSS project (EuroHPC-01-2019, G.A. n. 955648), The European PILOT project (EuroHPC-JU, G.A. n. 101034126), and by CINECA Italian national supercomputing center (ISCRA cod. HP10CHRZE7).

References

1. Taylor, R., Porto, F., Cui, C., Wadadekar, Y., Malkov, O.: Big data research infrastructure collaboration toward the SKA (BRICSKA). An. Acad. Bras. Cienc. **93** (2021)
2. Reina, G.A., Gruzdev, A., Foley, P., Perepelkina, O., Sharma, M., Davidyuk, I., et al.: Openfl: An open-source framework for federated learning. CoRR abs/2105.06413 (2021)
3. Beutel, D.J., Topal, T., Mathur, A., Qiu, X., Parcollet, T., Lane, N.D.: Flower: a friendly federated learning research framework. CoRR abs/2007.14390 (2020)
4. McMahan, B., Moore, E., Ramage, D., Hampson, S., y Arcas, B.A.: Communication-efficient learning of deep networks from decentralized data. In: Proc. of the 20th Intl. Conference on Artificial Intelligence and Statistics, AISTATS 2017. Proc. of Machine Learning Research, vol. 54, pp. 1273–1282. PMLR, Fort Lauderdale, FL, USA (2017)
5. Polato, M., Esposito, R., Aldinucci, M.: Boosting the federation: Cross-silo federated learning without gradient descent. In: Intl. Joint Conference on Neural Networks, IJCNN 2022, Padua, Italy, 2022. IEEE, New York (2022)
6. Kairouz, P., McMahan, H.B., Avent, B., Bellet, A., Bennis, M., et al.: Advances and open problems in federated learning. Found. Trends Mach. Learn. **14**, 1 (2021)
7. Shoham, N., Avidor, T., Keren, A., Israel, N., Benditkis, D., et al.: Overcoming forgetting in federated learning on non-iid data. CoRR abs/1910.07796 (2019)
8. Casella, B., Esposito, R., Cavazzoni, C., Aldinucci, M.: Benchmarking FedAvg and FedCurv for image classification tasks. In: Proceedings of the 1st Italian Conference on Big Data and Data Science, ITADATA 2022, 2022. CEUR Workshop Proceedings, Aachen (2022)

[2] https://github.com/alpha-unito/streamflow-fl.

9. Colonnelli, I., Cantalupo, B., Esposito, R., Pennisi, M., Spampinato, C., Aldinucci, M.: HPC application cloudification: the StreamFlow toolkit. In: PARMA-DITAM Workshop 2021. Open Access Series in Informatics (OASIcs), vol. 88, pp. 5–1513. Dagstuhl, Germany (2021)
10. Simonyan, K., Zisserman, A.: Very deep convolutional networks for large-scale image recognition. In: 3rd Intl. Conference on Learning Representations, ICLR 2015, San Diego, CA, USA, 2015, Conference Track Proceedings (2015)
11. Wu, Y., He, K.: Group normalization. In: Computer Vision - ECCV 2018 - 15th European Conference, Munich, Germany, Proceedings, Part XIII. Lecture Notes in Computer Science, vol. 11217, pp. 3–19. Springer, Heidelberg (2018)
12. LeCun, Y., Bottou, L., Bengio, Y., Haffner, P.: Gradient-based learning applied to document recognition. Proc. IEEE **86**(11), 2278–2324 (1998)
13. Netzer, Y., Wang, T., Coates, A., Bissacco, A., Wu, B., Ng, A.Y.: Reading digits in natural images with unsupervised feature learning. In: NIPS Workshop on Deep Learning and Unsupervised Feature Learning (2011)
14. Aldinucci, M., Rabellino, S., Pironti, M., Spiga, F., Viviani, P., et al.: HPC4AI, an AI-ondemand federated platform endeavour. In: ACM Computing Frontiers (2018)
15. Casella, B., Chisari, A.B., Battiato, S., Giuffrida, M.V.: Transfer learning via test-time neural networks aggregation. In: Proceedings of the 17th International Joint Conference on Computer Vision, Imaging and Computer Graphics Theory and Applications, VISIGRAPP 2022, Volume 5: VISAPP, 2022. SCITEPRESS, Setúbal (2022)
16. Crusoe, M.R., Abeln, S., Iosup, A., Amstutz, P., Chilton, J., et al.: Methods included: standardizing computational reuse and portability with the common workflow language. Commun. ACM **65**(6), 54–63 (2022)
17. Colonnelli, I., Cantalupo, B., Merelli, I., Aldinucci, M.: StreamFlow: cross-breeding cloud with HPC. IEEE Trans. Emerg. Top. Comput. **9**(4), 1723–1737 (2021)
18. Colonnelli, I., Aldinucci, M., Cantalupo, B., Padovani, L., Rabellino, S., et al.: Distributed workflows with Jupyter. Future Gener. Comput. Syst. **128**, 282–298 (2022)

Integration and Deployment of Model Serving Framework at Production Scale

Francesco Caronte and Eva Sciacca

Abstract With the help of machine learning systems, we can examine data, learn from that data, and make decisions. Nowadays machine learning models have become more relevant to various use cases, but to manage them, especially if numerous, is a challenging task. For this reason, several **Machine Learning Operations (MLOps)** tools have been developed. These tools are the main platforms, hosting the full machine learning process lifecycle, starting with data management, and ending with model versioning and deployment.

1 Introduction

NEANIAS[1] (Novel EOSC Services for Emerging Atmosphere, Underwater & Space Challenges) is an ambitious project that comprehensively addresses the challenges set out in the 'Roadmap for EOSC' foreseen actions. NEANIAS drives the co-design and delivery of innovative thematic services, derived from state-of-the-art research assets and practices in three major sectors: Underwater research, Atmospheric research and Space research.

The **Machine learning core services** are composed by a JupyterHub instance (C3.1 named AI-Gateway[2]) that enables different profile servers to: (1) serve machine learning models in production environment (C3.2); (2) enable a deep

[1] NEANIAS Project web page: https://www.neanias.eu/.

[2] The AI-Gateway is available at: https://ai-gateway.neanias.eu/.

F. Caronte (✉)
Altecspace SPA, Torino, Italy
e-mail: francesco.caronte@altecspace.it

E. Sciacca
INAF, Osservatorio Astrofisico di Catania, Catania, Italy
e-mail: eva.sciacca@inaf.it

F. Bufano et al. (eds.), *Machine Learning for Astrophysics*, Astrophysics and Space Science Proceedings 60, https://doi.org/10.1007/978-3-031-34167-0_40

learning training framework using a Horovod cluster (C3.3); and (3) perform distributed calculations using Apache Spark (C3.3).

All these services, combined and integrated, are used for astrophysics use cases. In particular, two environments were integrated in the AI-Gateway to classify radio astronomical sources, galaxies and image artefacts. These are based on two deep learning models tailored to **object detection** (based on maskRCNN) and **semantic segmentation** (based on Tiramisu) and trained on radio dataset [2] from the Square Kilometre Array (SKA) precursors. The services run on the Kubernetes container orchestration platform deployed on top of the OpenStack cloud infrastructure provided and maintained by GARR.[3]

2 Model Serving

Model serving is a platform that simplifies ML models deployment and enables to serve models at production scale in minutes. Model serving provides different libraries to **package** the model, **serving** it offline in the development environment or online using a Kubernetes cluster. An overview of the architecture is depicted in Fig. 1.

The model is served using flask and supports micro-batching mechanisms to perform prediction, and it automatically generates a deployment endpoint signed

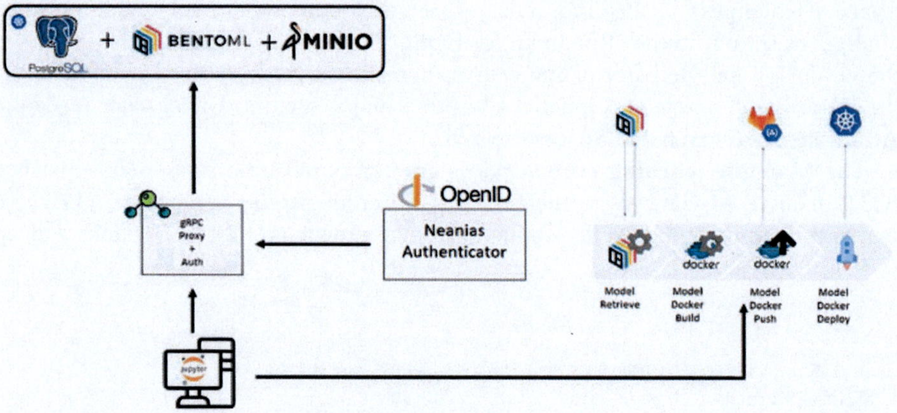

Fig. 1 Architecture of model serving using BentoML

[3] GARR Cloud Platform: https://cloud.garr.it/.

with authorized SSL. Model serving is implemented with the following technology stack:

- **BentoML**: an open platform that simplifies ML model deployment and enables to serve models at production scale[4] in minutes. It uses MinIO as High Performance Object Storage to store the artifacts and Yatai Server as backend
- **gRPC Proxy**: a proxy between C3.1 (JupyterHub) and Yatai service that permits to authorize and authenticate client requests
- **GitLab CI/CD**: used to deploy or destroy the machine learning model on the Kubernetes platform
- **Kubernetes**: an open-source system for automating deployment, scaling and management of the aforementioned containerized applications.

With the Model Serving solution[5] the Data Scientist can create different instances of the same model and deploy all the models on the Kubernetes cluster and run the prediction using the defined model API. Different type of data structures can be employed or, alternatively, a simple path contains the data to use for prediction. WeDav webservice provided by NextCloud is used to mount directories using the WebDav protocol.

The model serving have correctly integrated the Tiramisu model [1]. Tiramisu introduces a variation in the U-Net architecture, by employing a sequence of DenseNet blocks, rather than standard convolutional blocks in both downsampling (for feature extraction) and upsampling (for output generation) paths, with skip connections. This contributes to making the network require less parameters, as it reuses feature maps from earlier layers, freeing deeper layers from the need to learn redundant features. The model deployed give in input the path directory that contains radio astronomical images of our galaxy and the model save the prediction in an output directory returned by the prediction API.

3 Conclusion

The Model Serving can provide an easy way to turn the ML model into production-ready API endpoint and permits to standardize model packaging and ML service definition to streamline deployment, deploy and operate ML service workload at scale on Kubernetes using an automatic GitLab CI.

Acknowledgments The research has received funding from the EC Horizon 2020 grant agreement No. 863448 (NEANIAS).

---

[4] BentoML https://www.bentoml.ai/.

[5] Model Serving Framework documentation available at: https://docs.neanias.eu/projects/c3-2-modelserving/en/latest/.

References

1. Pino, C., Sortino, R., Sciacca, E., Riggi, S., Spampinato, C.: Semantic segmentation of radio-astronomical images. In: International Workshop on Artificial Intelligence and Pattern Recognition, pp. 393–403. Springer (2021)
2. Sortino, R., Magro, D., Sciacca, E., Riggi, S., Fiameni, G.: Radio astronomical images object detection and segmentation. In: A benchmark on deep learning methods. pp. 1–39. Astronomy (2023)

Predictive Maintenance for Array of Cherenkov Telescopes

Salvatore Gambadoro, Salvatore Cavalieri, Alessandro Costa, Diego D'Urso, Federico Incardona, Kevin Munari, and Pietro Bruno

Abstract A predictive maintenance (PdM) approach is considered among the best solution for the maintenance of equipment. The paper aims at presenting an approach to the realization of a predictive maintenance for the array of Cherenkov telescopes. The application of PdM is able to reduce array downtimes, to increase the longevity of telescope components, and to reduce costs due to unscheduled maintenance activities.

Keywords Predictive maintenance · Machine learning · Cherenkov Telescope Array

1 Introduction

In the astronomical field, availability of running instruments and, more specifically, telescopes that are always operational, or for which the idle time is reduced as much as possible, is fundamental for the success of experiments. In recent years, thanks to the spread of the Internet of Things (IoT) [1], telescopes are equipped with a multitude of sensors, scattered throughout the instrumentation, able to provide a large volume of telemetric data. These data are fundamental for maintenance purposes, because they allow us to apply a new approach developed

S. Gambadoro (✉)
Department of Electrical Electronic and Computer Engineering, University of Catania, Catania, Italy

INAF, Osservatorio Astrofisico di Catania, Catania, Italy
e-mail: salvatore.gambadoro@phd.unict.it

S. Cavalieri · D. D'Urso
Department of Electrical Electronic and Computer Engineering, University of Catania, Catania, Italy

A. Costa · F. Incardona · K. Munari · P. Bruno
INAF, Osservatorio Astrofisico di Catania, Catania, Italy

F. Bufano et al. (eds.), *Machine Learning for Astrophysics*, Astrophysics and Space Science Proceedings 60, https://doi.org/10.1007/978-3-031-34167-0_41

in recent years called Predictive Maintenance (PdM) [2]. PdM improves the overall productivity and efficiency of your systems and allows you to predict failures early enough to prevent complete system failure. The idea behind PdM is to use a data-driven approach that uses the data gathered by the system to predict potential major failures. With appropriate processing and use of artificial intelligence (AI) algorithms, it is possible to predict whether a component will fail. Our idea was to create a PdM system for an astronomical telescope; in particular, it was considered the Cherenckov Telescope Array (CTA) [3, 4].

2 Description of the Scenario

CTA is the largest and most advanced ground based facility for the detection of very high energy electromagnetic radiation, from 20 GeV to 300 TeV. The telescope used for the case study is a prototype of a Small Size Telescope Cherenkov located in Catania at the Serra La Nave observatory [4]. The first research step was focused on a reliability analysis of the system, based on the use of Fault Tree Analysis (FTA) [5]. FTA pointed out that the components that result to most likely have failures are the motors that allow the telescope to lift and rotate. The available dataset collected was a 5-year time series (January 2017–February 2022). Unfortunately, a small amount of information was available about the occurrence of faults and maintenance operations of each set of relevant values for a particular timestamp. For this reason, in the absence of labels about fault conditions, the idea was to apply unsupervised learning techniques, realizing the PdM through a process aimed to point out the anomaly detection, where the anomaly is made up by the failure condition of a motor.

3 Results

Several algorithms have been used for the generation of the machine learning-based model of our interest, to highlight the best performance. Unsupervised models based on Clustering-Based Local Outliers (CBLO) [6] and Principal Component Analysis (PCA) [7] allowed to achieve the best performances. According to the analysis performed on the data set at our disposal, the anomaly points are 874, but from the information in our possession, only 102 refer to real system failures. Many of the anomalies found in the dataset did not led to a real breakdown of the system; the 102 anomalies found did led to a real breakdown in the next future. Considering these 102 anomalies, CBLO and PCA models were able to detect 93 out of them, indicating a 91% success rate. Figure 1 shows some of the anomalies correctly recognised by the models; these anomalies are shown by blue points and occurred in November 2021; they led to a breakdown of the system occurred in February 2022; the brake on the telescope failed and the telescope was unusable for a long

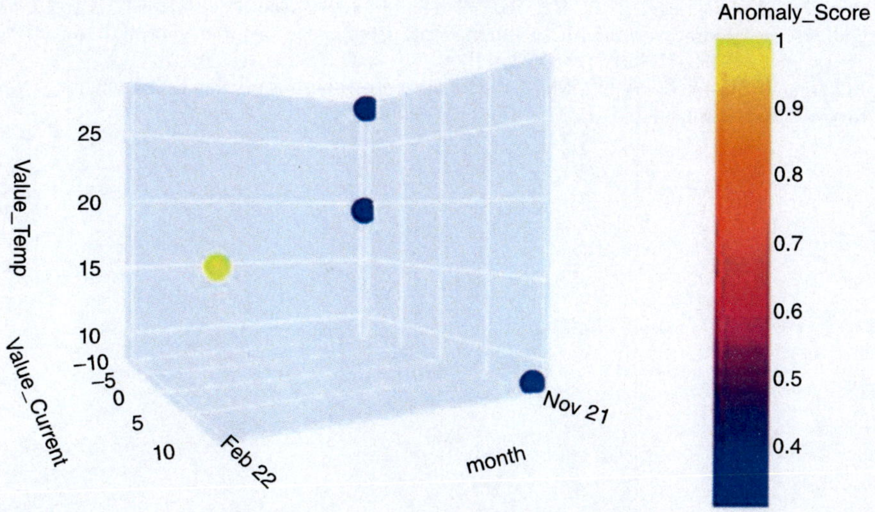

Fig. 1 The classification results. Three anomalies detected in November and the real breakdown in February

time. Detection of the anomalies in November 2021 could allow the realisation of a maintenance procedure, preventing the telescope's engine from breaking, and reducing telescope downtime.

4 Conclusions

The aim of the proposed research is to propose a PdM system for telescopes, using a machine learning model. This model may allow to minimize downtime, to increase the longevity of telescope sub-components, and to reduce maintenance costs.

References

1. Kumar, S., Tiwari, P., Zymbler, M.: Internet of Things is a revolutionary approach for future technology enhancement: a review. J. Big Data. **6**(111) (2019)
2. Mobley, R.K.: An Introduction to Predictive Maintenance. Elsevier (2002)
3. Acharya, B., et al.: Introducing the cta concept. Astropart. Phys. **43**, 3–18 (2013)
4. Maccarone, M.C.: ASTRI for the Cherenkov Telescope Array. Proceedings 35th International Cosmic Ray Conference (2017)
5. Ericson II, C.A.: Introduction to Fault Tree Analysis, pp. 4–13. Safety Solutions LLC (2014)

6. Breunig, M.M., Kriegel, H.P., Ng, R.T., Sander, J.: LOF: Identifying Density-Based Local Outliers. Proceedings of SIGMOD International Conference on Management Data, Dallas, Texas, pp. 427–438. (2000)
7. Rokhlin, V., Szlam, A., Tygert, M.: A randomized algorithm for principal component analysis. SIAM J. Matrix Anal. Appl. **31**(3), 1100–1124 (2009)

Index

F. Bufano et al. (eds.), *Machine Learning for Astrophysics*, Astrophysics and Space Science Proceedings 60, https://doi.org/10.1007/978-3-031-34167-0